Women of Piracy

Drawing from an interdisciplinary body of research and data, *Women of Piracy* employs a criminological lens to explore how women have been involved in, and impacted by, maritime piracy operations from the sixteenth century to present-day piracy off the coast of Somalia.

The book challenges and resists popular understandings of women as peripheral to the criminal enterprise of piracy by presenting and analyzing their roles and experiences as victims, perpetrators, and criminal justice actors, showing that women have been, and continue to be, central figures in maritime piracy. Unfolding in four chapters, Chapter 1 sets the context by providing readers with a history of gender and the female imaginary. Chapter 2 focuses on the gendered social organization of piracy operations, discussing how and why the roles and responsibilities associated with this gendered labor have emerged, persisted, evolved, and/or ceased over time, as well as considering which roles and responsibilities appear to be context specific and which seem to transgress geographical locations. Chapter 3 introduces and analyzes the women of piracy-adjacent trafficking and their victims. Lastly, Chapter 4 explores how women have (or have not) been brought to justice for their participation in crimes of piracy as well as the roles of women in efforts to combat piracy. The overarching objective is to ignite a broader discussion about the various cultural, social, historical, and economic forces that create opportunities for women to participate in maritime piracy and counter-piracy, why women continue to be invisible figures of piracy, and what implications this has for how we study, police, and bring pirates to justice.

The first criminologically grounded, global study exploring the continuity and evolution of women in maritime piracy, this book will be of great interest to students and scholars of criminology, gender, feminist studies, international relations, anthropology, history, and political geography. It will also be useful to maritime and law enforcement professionals.

Brittany VandeBerg is an Associate Professor in the Department of Criminology and Criminal Justice at the University of Alabama, USA. She is an interdisciplinary scholar whose research aims to better understand the relationship between territory, violence, and law. Her areas of expertise include maritime piracy, legal geographies, gender and crime, and criminal justice reform, particularly in sub-Saharan Africa. VandeBerg is a former consultant with the United Nations Office on Drugs and Crime (UNODC) Counter Piracy Programme (since renamed the Global Maritime Crime Programme) and the Food and Agriculture Organization (FAO) of the United Nations Somalia Fisheries Sector.

Feminist Criminology

This series is a natural home for Criminology research with a Feminist Studies focus. Bringing together original, innovative and topical books that showcase cutting edge theory and empirical research, it is a focal point around which the field can continue to develop and flourish. The series is broad in scope, in recognition of the diverse nature of research that is undertaken relating to Feminist Studies, Crime and Criminal justice.

Gender, Homicide and the Politics of Responsibility
Ashlee Gore

Women, Crime and Justice in Context: Contemporary Perspectives in Feminist Criminology from Australia and New Zealand
Edited by: Anita Gibbs and Fairleigh Evelyn Gilmour

The Social Exclusion of Incarcerated Women with Cognitive Disabilities: Shut Out, Shut In
Julie-Anne Toohey

Women of Piracy
Brittany VandeBerg

Women of Piracy

Brittany VandeBerg

LONDON AND NEW YORK

First published 2023
by Routledge
4 Park Square, Milton Park, Abingdon, Oxon OX14 4RN

and by Routledge
605 Third Avenue, New York, NY 10158

Routledge is an imprint of the Taylor & Francis Group, an informa business

British Library Cataloguing-in-Publication Data
A catalogue record for this book is available from the British Library

ISBN: 978-1-032-11912-0 (hbk)
ISBN: 978-1-032-12571-8 (pbk)
ISBN: 978-1-003-22520-1 (ebk)

DOI: 10.4324/9781003225201

Typeset in Times New Roman
by Deanta Global Publishing Services, Chennai, India

For Cadence and Callum

Contents

Acknowledgements

This book would not have been possible without the wide-ranging support from Susan Dewey whose insights, ideas, enthusiasm, and critical reading of the manuscript undoubtedly shaped the final draft. I am deeply grateful for her friendship and unwavering emotional and intellectual support. I am also especially grateful to my brilliant and inspirational feminist colleagues of the Organized Crime Group, Felia Allum, Patricia Figueroa, Elaine Carey, and Deborah Bonello for our continuing conversations about women in international organized crime. Lastly, I remain grateful to George Thompson, publisher-in-residence at the University of Alabama, for encouraging me along the way as this project evolved from a pre-Covid proposal to a post-Covid published book.

List of Acronyms and Abbreviations

AISE	Italian Intelligence Service
CVE	Countering Violent Extremism
GoG	Gulf of Guinea
UNCLOS	United Nations Convention on the Law of the Sea
UNDP	United Nations Development Programs
UNODC	United Nations Office on Drugs and Crime

Introduction

It was late morning in Garowe, the arid northeastern capital of Puntland, and the oven that is June in the Horn of Africa was already in full force as my Norwegian colleague and I arrived to monitor the progress of a prison construction project facilitated by the United Nations Office on Drugs and Crime (UNODC) Counter Piracy Programme. It was the largest prison project ever undertaken by UNODC—a 500-bed detention unit, officer housing, and training facility for training Somali prison staff in correctional principles, policies, and practices. UNODC considered the prison the shining exemplar of its fledgling regional counter-piracy project's goal to provide humane and secure detention conditions in line with international human rights standards. These standards, which reflect Western European standards for confinement, guaranteed its future residents—Somali men convicted of piracy—balanced diets, potable water, and well-constructed housing. The terrible irony is that if its residents could support a family while incarcerated, the prison would fill the vacuum of need that drove these young men to piracy in the first place.

The building contractor proudly led us around the construction site, pointing out the newly dug bore hole and each future cell block's footprint. The construction workers took a break from the heat to greet us as we shielded our eyes from the sand's perennial scuttle across the barren landscape, its low scrub brush the only vegetation in sight. Our Somali interpreter beckoned us back to our vehicle, where a representative from then president Farole's office had summoned us by phone to a downtown government building. All United Nations personnel in Garowe were required to be present or face expulsion from Somalia, and so we hurried into town.

We were ushered into a large room upon our arrival, where ten Somali elders and government officials sat at tables talking loudly with one another. A hush fell over the room when they saw me walk in. I nervously fidgeted with my scarf thinking, perhaps, too much of my hair was exposed, a perpetual concern for women in the Horn of Africa. Our interpreter introduced

DOI: 10.4324/9781003225201-1

us in Somali as UNODC representatives. Utter silence. My colleague tried to break the tension by waving his hand and saying "hello" in his singsong Norwegian accent. A man sitting toward the front of the room finally stood up and declared in English, "A woman? Working on piracy?!". He burst into laughter and the others boisterously joined in with misogynistic comments in Somali and English about what they regarded as the absurdity of the situation.

All of my insecurities rushed to the surface. I had spent nearly a decade studying gender mainstreaming in development, yet here I was doubting that a small town girl from rural Wisconsin had anything important to say. Despite being fully covered from head to toe, I felt more exposed than ever. I tried my best to conceal the disappointment and embarrassment on my face and stared silently at these howling men for a few moments before their laughter wound down with the creeping realization that, in fact, they had no choice but to listen to me. They did not know it yet, but I was just one of the first of many female UNODC staff members who would work on piracy over the next few years and that ten years on the entire regional program would be led by a woman.

I vividly remember that meeting as my first experience with piracy's gendered realities. If women had no place in popular understandings of maritime piracy in Somalia, they certainly did not shape counter-piracy policy and practice. Until I arrived, that is. Through endless phone calls, emails, and trips to Somalia, I eventually earned the respect, confidence, and even the cautious friendship of many men who heckled me in the room that day. I try not to think about all the other work I could have accomplished during the time sexist attitudes forced me to waste, convincing Somali men that I was, in fact, a competent professional.

Almost a decade after that meeting took place, this book repositions women from the margins to the center of discussions about maritime piracy, particularly contemporary piracy, and inspire further scholarly explorations into the diversity of women's experiences and roles in maritime piracy. Maritime piracy is one of the oldest transnational crimes, touching nearly every region of the world, yet piracy has predominantly been studied, policed, and prosecuted through a gendered and, arguably, Western gaze. Men take center stage in popular cultural and scholarly accounts as piracy's real actors—the leaders, the crewmembers, and the masterminds. Women, when featured, are either helpless figures victimized by pirates or deviant women who reject their femininity to join the pirate ranks. These accounts depict piratical women as inherently unnatural products of tough life circumstances.

Feminist criminologists recognize how women's varied experiences with crime and justice are based on race, gender, nationality, socioeconomic status, and a multitude of other factors. Despite the historical tendency to explain women's and men's lawbreaking crime by using theories and frameworks designed by men to understand other men (Irwin & Chesney-Lind, 2008), feminist criminological engagements with women offenders push back and underscore the construction and performance of gender in the broader context of gender, class, and racial privilege (Chesney-Lind & Morash, 2013). Challenging popular understandings of gender as something "done" in relation to crime allows us to explore how multiple systems of oppression interact and produce contexts in which women's violence and lawbreaking make sense, particularly when contextualized within women's need to survive (Chesney Lind & Jones, 2010). I accordingly take a gendered approach to piracy to unite two complex imaginaries—women and pirates—and advance a more empirically grounded and conceptually rich analysis of piracy.

Piracy as We Know It

To better understand how we have come to *know* piracy, we must first identify the "authorized knowers" of piracy, the knowledge claims they make, and the potential limitations of current piracy scholarship. Authorized knowers are those dominant cultural institutions, such as universities and governments, that construct ideal knowers whose words and ideas automatically receive these powerful institutions' subsequent acceptance and endorsement as expert knowledge (Smith, 1974; Moore & Muller, 1999; Hunter, 2002; Maton, 2013). For the past two decades, criminologists have engaged with and problematized the notion of expert knowledge and authorized knowers to critique several of the underlying foundations of criminological thought such as the causes of "crime" and "punishable subjects" (Smart, 1995; Snider, 2003). Yet, beyond the pathbreaking work of feminist criminologists, criminology as a field generally fails to critically engage with how the discipline, and its foundational tenets, traditionally embraces male scientists as authorized knowers. The struggles continues against conceptualizing female offenders as morally corrupt and in need of refinement or biologically flawed and deserving of harsher punishments for their sins against social norms.

Piracy's authorized knowers approach the subject from a diversity of disciplines, including history, political science, international relations, economics, geography, anthropology, and criminology. Whereas most of the data informing historical studies of piracy is derived from piracy trial transcripts, official government papers, and captain's journal, all of which were produced by men and filtered through the male gaze (Turley, 1999), contemporary data collection includes unique challenges. Despite a new wave of "techno-optimism" resulting from the use of satellites and unmanned aerial vessels, monitoring maritime activities has long been a challenge and illegal activities continue to flourish on the ocean (Daxecker & Prins, 2013; Nyman, 2019). There is a lack of clear, consistent, and transparent reporting on contemporary piracy attacks (Hurlburt, 2013), and most data is derived from a self-reporting system. With many shipping companies hesitant to report piracy attacks so as not to drive up insurance rates and spook investors, and fishing companies involved in illegal fishing aiming to avoid self-incrimination, there is a serious underreporting of piracy incidents around the world (Vagg, 1995; Barrios, 2013). Even anthropologists and other ethnographic researchers who might initially be thrilled to conduct fieldwork with pirates would quickly recoil at piracy's quotidian realities: hostage-taking, sexual assault, torture, extortion, and murder. To my knowledge, my first book is the sole example of institutional ethnography on counter-piracy (Gilmer, 2014).

Due to piracy's international nature, the UNODC's Global Maritime Programme (formerly the Counter Piracy Programme) continues to spearhead the collection and distribution of contemporary piracy data. Bueger (2015) brilliantly explores the international epistemic infrastructures, practices, and laboratories that produce "archetypes" of contemporary piracy knowledge for the United Nations, which is then filtered through various global outlets. These archetypes involve quantification by providing evidence in the form of reports, interpretation of local knowledge, and consulting a network of special advisors. The data generated through the United Nation's epistemic practices, thus, serves as the raw material from which practitioners, policymakers, and academics come to produce the *knowns* about piracy. Very few of these *knowns* pertain to women.

UNODC data informs a vast portion of the extensive body of contemporary piracy studies addressing maritime piracy in coastal waters bordering Somalia, Southeast Asia, and the Gulf of Guinea (GoG). These studies fall into three primary categories pertaining to piracy: (a) its causes, organizational structures, and practices; (b) institutional responses; and (c) attempts to historicize and deconstruct piracy's geopolitics (Bueger, 2014). Most of this literature is very practitioner oriented in its focus on informing and/or evaluating counter-piracy programming efforts. For example, article titles often include phrases such as anatomy of an attack, combating piracy, fighting piracy, suppressing maritime piracy, "military capacity and piracy, and piracy and maritime security. One cannot understate the importance of these studies in informing policy and best practices. However, unlike studies utilizing a historical, anthropological, geographical, and sociological approach that attend to the social complexities and political dynamics of piracy through the ages (Leeson, 2009; Dawdy, 2011; Dawdy & Bonni, 2012; Dent, 2012; Dua, 2019), these studies rather nervously shy away from engaging with piracy as a sociological phenomenon linked to broader questions of sovereignty, empire, morality, and deviance.

Criminologists, although slow to engage with the ocean as a social space, have begun to identify some of the limitations of existing piracy research, and thus potential avenues of inquiries that can move understandings of the criminal enterprise forward into exciting new directions. First, while being careful not to overstate continuities between historical and contemporary piracy, criminologists suggest following the lead of historians and delving beneath the "public appearance" of piracy to better understand the "private" (or unseen) aspects of piracy to include sociocultural organization, forms of recruitment, relations to participants in land-based networks, officials, police, and more (Ong-Webb, 2006). Second, contemporary piracy research stands to benefit from reevaluating the politics behind and transformation of definitions and typologies of piracy (Twyman-Ghoshal, 2014). Because crime is a social construct, the way societies imagine, define, understand, and engage with crime is fluid and constantly (re)constituted in relation to broader social changes. Third, taking cues from situational crime prevention studies, it is important to acknowledge piracy's spatiality. Contemporary piracy attacks occur in geographical clusters involving an assemblage of actors and institutions, yet most piracy victims and offenders are from the Global South (Marchione & Johnson, 2013; Townsley et al., 2016) and fundamentally absent from the piracy knowledge production process. Lastly, most scholarly research on women in piracy stops at the twentieth century, perpetuating assumptions that women are "unlikely to play an active part in what is still a man's world" (Stanley et al., 1995, p. 254).

Seeing beyond the Masculinized Pirate Mirage

A handful of delightfully entertaining books excitedly attend to the stories of well-known women pirates (see Klaussmann et al., 1997; Duncombe, 2017, 2019), but very few academic studies of piracy consider gender as a central component of analysis (Gilmer, 2017). Among these are a few influential historical studies (see Creighton & Norling, 1996; Cordingly, 2002; Appleby, 2013) that establish the groundwork for a broader project of understanding piratical women's agency within the broader context of economic insecurity, politics, and gender ideologies. Cultural criminologists have been at the forefront of parsing out the collective meaning of crime, criminal, deviance, and the illicit as a constant negotiation between individuals, structures, and society (Young, 2007). Understanding crime and criminals, then, requires understanding culture (Hayward, 2004). Piracy is traditionally understood as a "male-dominated" cultural activity in which women are an afterthought. This book reimagines the possibilities that piratical women engage(d) in piracy-related activity in ways that do not simply mimic those engaged in by men, but rather in ways that could be considered "women-dominated" practices and events.

Generally, historical and contemporary piracy studies have masculinized maritime piracy—what I call producing the *masculinized pirate mirage*. The masculinized pirate mirage is the product of narrowly scoped policy discourse, academic scholarship, media coverage, entertainment, and media that prioritizes the study and promotion of men as *the* subjects of piracy. This is evidenced through a textual and visual bias at the international scale that, like the UNODC archetypes previously discussed, continues to identify the "offenders" of maritime piracy as generally young men raised in desperate and disorganized society, willing to risk their lives for the slightest chance of something better. These authorized knowers of piracy regularly provide one-hundred-plus page documents that invisibilize women. This bias also occurs among news outlets at the state and local level, which are then taken up and promoted globally, such as the now iconic photo of a Somali man wrapped in a *macawis* (customary male sarong) wielding an AK-47 as he stands on the Somali shoreline staring off into the distant waters of the Indian Ocean.

The constant barrage of these images reinforces understandings of piracy as a man's crime carried out by hypermasculine, aggressive offenders while simultaneously hindering efforts to understand the crime from a sociological perspective and formulate appropriate responses based on those understandings. The material effect of framing of pirates as dangerous, rogue, *male* maritime actors is the generation of moral panic allowing for the excessive militarization of ocean space and the legitimization of state violence in the name of crime prevention and suppression (Collins, 2012; Gilmer, 2014; Dua, 2018). Paradoxically, pirates and their associates also benefit from the masculinized pirate mirage in that women, who are a large part of piracy's operational networks, operate behind the scenes in roles where they earn less money, prestige, and status than their male peers. Thus, the *portrayal* and *performance* of piracy as a man's crime establishes subjects and spaces of intervention that works to the advantage of many stakeholders, revealing a much messier, hazier, dynamic assemblage of actors and geographies that has long been reliant on the politics of gender and the participation of women.

Objectives and Frames

The overarching objective of this book is to envision an alternative to the masculinized pirate mirage by employing an interdisciplinary, multiscalar, and reflexive feminist criminological approach to reexamine women's involvement in maritime piracy. This approach has three objectives. First, it allows for the (re)presentation of piratical women as heterogeneous subjects who actively participate in the creation, reinforcement, and challenging of various gendered labor roles of maritime piracy operations. Second, it discusses how and why the roles and responsibilities associated with this gendered labor emerged, persisted, evolved, and/or ceased over time. Third, it considers which roles and responsibilities appear to be context specific and which seem to transgress geographical locations. Through these objectives, I hope to ignite a broader discussion about the various cultural, social, historical, and economic forces that create opportunities for women to participate in maritime piracy and counter-piracy, why women continue to be invisible figures of piracy, and the implications this has for how we study, police, and bring pirates to justice.

I embed these objectives within a framework that first acknowledges how maritime piracy is a dynamic social phenomenon historically tied to efforts of state building and personal greed. Although continuities exist between historical and contemporary piracy, the definition of piracy and its operational aspects continue to change and evolve. Similarly, *where* piracy takes place remains contested and politicized. As such, I work from the current definition of piracy as defined by Article 101 of the United Nations Convention on the Law of the Sea (UNCLOS). Piracy consists of any of the following acts:

- a. any illegal acts of violence or detention, or any act of depredation, committed for private ends by the crew or the passengers of a private ship or a private aircraft, and directed:
  - i. on the high seas, against another ship or aircraft, or against persons or property on board such ship or aircraft;
  - ii. against a ship, aircraft, persons or property in a place outside the jurisdiction of any State;
- b. any act of voluntary participation in the operation of a ship or of an aircraft with knowledge of facts making it a pirate ship or aircraft;
- c. any act of inciting or of intentionally facilitating an act described in subparagraph (a) or (b). (United Nations, 1982/2022)

While this definition is the closest possible to an "agreed upon" international definition of maritime piracy, I fully recognize its limitations and the geopolitics of Western bias underlying its adoption. However, it is a useful starting point from which to begin to understand how crime as a social construct becomes defined and codified in laws that may not match the realities on the ground, and therefore, demands ongoing reassessment.

My analytical framework approaches and analyzes piracy as a form of organized crime. The United Nations Convention against Transnational Organized Crime defines an organized criminal group as

> A structured group of three or more persons, existing for a period of time and acting in concert with the aim of committing one or more serious crimes or offences established in accordance with this Convention, in order to obtain, directly or indirectly, a financial or other material benefit.
>
> (United Nations, 2004: 5)

However, I prefer Abadinsky's (1985/2012) broad conceptualization of organized crime as consisting of the following attributes: absence of political goals; hierarchal, limited, or exclusive membership; constitutes a unique subculture; perpetuates itself; exhibits a willingness to use illegal violence; is monopolistic; and is governed by explicit rules and regulations (p. 3). This broader conceptualization allows flexibility for expanding our understanding of who and what constitutes organized crime. As the book unfolds, it is my hope that the reader will see the need for an equally flexible revised definition of piracy.

Finally, my analytical framework centers women in piracy studies as part of a broader project to abolish the unproductive characterization of maritime crimes as committed on the *edges* of our consciousness and society. Geographers have been at the forefront of critiquing understandings of the social and cultural worlds of the sea as somehow inferior and marginal to land, when the sea is nonetheless a space "integral to the workings of the world as we currently know it" (Peters, 2010, p. 1260; see also Steinberg, 2001; Lambert et al., 2006; Peters, 2016). It is impossible to separate the social worlds of the sea wholly from the social worlds of land, including crime. Thus, it is important to investigate and challenge piracy-related policies, practices, and imaginations that remain fixated on the illicit activities at sea while giving those who reside on shore a pass by downplaying their importance (Hesse, 2013).

While my intent is not to fully dismantle existing scholarship on piracy, I hope to broaden and refine the Western male "gaze" to consider knowledge, representation, and empirical examples beyond "the west" through a criminological lens. I approached the analysis keeping in mind how (piratical) women are both discursively constructed and material subjects of their own history (Mohanty, 1988). Where possible, I attempted to avoid the analytical trap of assuming women to be victims of male violence, universal dependents, victims of colonial and development processes, or distilled representations of their religious ideologies. I examine empirical examples of piratical women as "situated" to attend to the unique geopolitical context of the women, their identities, and their actions (Aas, 2012).

Methods

This book is the culmination of the past ten years conducting research on piracy off the coast of Somalia. My journey began as a PhD student at the

University of Toronto when I was fortunate enough to land an internship opportunity with the UNODC Counter Piracy Programme (since renamed the Global Maritime Crime Programme) based in Nairobi, Kenya. From 2010 to 2012, I lived in Kenya and traveled back and forth to Canada and the United States while I fulfilled the requirements for my doctoral degree. During this time, my internship evolved into a consultancy position that enabled me to travel throughout Somalia and the broader East Africa region to witness firsthand the dangers, damage, and opportunities brought about by piracy off the Somali coast.

In my capacity as a consultant and a researcher, I conducted semi-structured interviews, held focus groups, led data analysis workshops to verify the results with local realities, watched and participated in the daily unfolding of counter-piracy policy and practice, first at UNODC and later with the UN Food and Agriculture Organisation Somalia Fisheries Sector. I acknowledge my role in many of the perceived programming shortcomings and failures, as well as my complicity in the broader counter-piracy assemblage that my scholarship continues to critique. As a white, Western woman employed by the United Nations, I am also keenly aware of the many privileges I was afforded as I moved throughout Somalia, Kenya, and the Seychelles with relative ease and security. I was granted unprecedented access to incarcerated pirates, earned the trust of high-ranking government officials, worked alongside key stakeholders and community organizers in Somalia, and befriended many foreign intelligence officers who were willing to share unclassified information with me. All these interactions helped shape my understanding of what was happening both off the coast and on the ground in Somalia.

Whereas many policymakers, practitioners, and academics packed up and moved on to the next hot issue when the threat of Somali piracy all but disappeared in 2013, I spent the past eight years immersing myself in interdisciplinary literature that all seemed to paint the same picture of what Somalia piracy is/was and who Somali pirates were/are. My former UNODC colleagues would joke, "Brit, nobody cares about piracy anymore". But I did, because the motivations of piracy, the calculated response to it, and the realities of its impact for various people and communities were far more complex than what was being captured in academic literature and policy reviews. The world was either hearing a carefully crafted Western, military perspective of piracy or an equally carefully crafted "in defense of Somalia" perspective from a handful of vocal pirates. Those two frames and their accompanying narratives neglected to capture the experiences of thousands of hostages, thousands of family members of perpetrators and victims, the regretful pirate, the corrupt politician, the career-driven United Nations worker, and more. Every time I returned to Somalia or Kenya to collect archival documents and touch base with those who, like me, found themselves caught in the thick of it when the piracy crisis was unfolding, I was reminded there is more of the story to be told.

My analysis puts primary data in conversation with secondary data produced by brilliant scholars working on the "fringes" of criminology and piracy studies. While researching and writing, I was greatly inspired by feminists pushing the boundaries of organized crime studies and gained a newfound appreciation for the sociocultural approach to piracy employed by historians and anthropologists. The vignettes included in the chapters feature composite characters based on fieldnotes collected during my time traveling around the Puntland and Somaliland regions of Somalia talking with women, community organizers, government officials, and counter-piracy staff. With these vignettes, I hope to bring to life the realities of contemporary piracy and inspire new female pirate imaginaries.

Throughout the book, I refrained from providing specific names and have changed some of the locations where events took place to protect the confidentiality of those who were generous enough to share their stories and insight with me. Somalia is a country with vast networks of familial and clan linkages, where everyone knows everyone and word of mouth travels faster than the internet. For this reason, I insist on preserving the confidentiality of my sources while still honoring their wish to have their lived experiences be part of the piracy dialogue.

Organization of the Book

The book unfolds in four chapters and a conclusion. Chapter 1 discusses how gendered assumptions about the sea have and continue to impact how maritime spaces, actors, and activities are studied and imagined, particularly piracy. The woman pirate imaginary—the fanciful image conjured up when one thinks of what a woman pirate looks like, how she acts, what she does, that is, both a product of and reinforced through literature, film, and even piracy studies—amplifies this false dichotomy through two seemingly opposing constructs. She is either normalized as a "victim" without agency or a fantastical "leader/fighter" who exercises too much agency in her willful disregard for others. Accordingly, the gendering of piracy is both a lived reality and a political project that reproduces popular imaginaries of how women experience and perform the violence of piracy.

Chapter 2 attempts to strike a fine balance between deepening understandings of female pirate leaders while attempting to avoid oversimplifying these women's complex lived realities. Knowledge is limited by the small body of academic literature on the topic and the recognition that most of these women's stories and lives were recounted and historicized by men. Without autobiographical evidence, interviews, or testimonies, these women's legacies are not entirely their own, and at times, speculative at best. Yet, this chapter utilizes these histories to glean women's pathways to pirate leadership, highlight the similarities they share with women involved in other forms of transnational crime, and theorize whether women pirate leaders can be understood as maritime *edgeworkers*—women who voluntarily participate in high-risk activities as a form of escape and resistance in ways that challenge and reshape the boundaries of gender, society, and the sea.

Chapter 3 discusses women's role in piracy within the global dialogue on human trafficking. Human trafficking priorities have taken various forms globally, including sex trafficking, labor trafficking and exploitation, organ trafficking, child trafficking and exploitation, child soldering, and bride trafficking (Dragiewicz, 2014). This chapter contributes an empirical example of ransom piracy in Somalia to highlight the interdependency between labor and sex trafficking and piracy not yet explored in either literature. As we will see, these forms of trafficking are piracy adjacent, rather than central to the operations of piracy, yet are nonetheless important sources of revenue for pirates. Trafficking in Somalia, as elsewhere in the world, is a gendered phenomenon, with men and women generally filling unique roles as offenders and victims of forced prostitution, domestic servitude, bonded labor, forced labor, and slavery.

Chapter 4 provides an overview of the history of punishments for piracy as contingent on where the crime was committed and prosecuted and the citizenship, age, and race of the perpetrators and victims. Gender also influenced the method and severity of punishment piracy. As understandings about women who offend change over time, so do understandings about what constitutes acceptable and effective punishments for their transgressions (Snider, 2003). This chapter explores how various approaches and techniques for punishing pirates—from hangings to jail sentences to amnesty—have been adapted or abandoned when bringing female pirates to justice. It also examines the instrumentalization of gender and motherhood by contemporary counter-piracy programming efforts to suppress piracy off the coast of Somalia to argue that punishment for piracy is a phenomenon as gendered as piracy itself.

The conclusion begins with a discussion that explores several important questions that feminist scholars continue to engage with as they advance understandings of women and crime. It then revisits what popular culture reveals about female pirates and beliefs regarding gender, sexuality, and race and the gendered glorification of crime. It reviews what the gendered social organization of piracy reveals about the nuances of broader social struggles and argues that historical and practitioner accounts of piracy present a limited understanding of the relationship between gender and crime in relation to piracy. The case study of piratical women also reveals the complex means by which women exercise agency from a limited menu of life choices. The chapter concludes with suggestions and hopes for moving forward.

References

Aas, K. F. (2012). 'The Earth is one but the world is not': Criminological theory and its geopolitical divisions. *Theoretical Criminology*, *16*(1), 5–20.

Abadinsky, H. (1985/2012). *Organized crime*. Boston: Cengage Learning.

Appleby, J. C. (2013). *Women and English piracy, 1540–1720: Partners and victims of crime*. Rochester, NY: Boydell Press.

Barrios, C. (2013). Fighting piracy in the Gulf of Guinea. *European Union Insfitute for Security Studies Brief.*

Bueger, C. (2014). Piracy studies: Academic responses to the return of an ancient menace. *Cooperation and Conflict*, *49*(3), 406–416.

Bueger, C. (2015). Making things known: Epistemic practices, the United Nations, and the translation of piracy. *International Political Sociology*, *9*(1), 1–18.

Chesney-Lind, M., & Jones, N. (Eds.). (2010). *Fighting for girls: New perspectives on gender and violence*. Albany: SUNY Press.

Chesney-Lind, M., & Morash, M. (2013). Transformative feminist criminology: A critical re-thinking of a discipline. *Critical Criminology*, *21*(3), 287–304.

Collins, V. E. (2012). Dangerous seas: Moral panic and the Somali pirate. *Australian and New Zealand Journal of Criminology*, *45*(1), 106–132.

Cordingly, D. (2002). *Seafaring women: Adventures of pirate queens, female stowaways, and sailors' wives*. New York: Random House Trade Paperbacks.

Creighton, M. S., & Norling, L. (Eds.). (1996). *Iron men, wooden women: Gender and seafaring in the Atlantic World, 1700–1920*. Baltimore: JHU Press.

Dawdy, S. L. (2011). Why pirates are back. *Annual Review of Law and Social Science*, *7*, 361–385.

Dawdy, S. L., & Bonni, J. (2012). Towards a general theory of piracy. Anthropological qaurterly, *83*(3), 673–699.

Daxecker, U., & Prins, B. (2013). Insurgents of the sea: Institutional and economic opportunities for maritime piracy. *Journal of Conflict Resolution*, *57*(6), 940–965.

Dent, A. S. (2012). Introduction: Understanding the war on piracy, or why we need more anthropology of pirates. *Anthropological Quarterly*, *85*(3), 659–672.

Dragiewicz, M. (2014). *Global human trafficking: Critical issues and contexts*. New York: Routledge.

Dua, J. (2018). Privateers and public ends: Piracy as global moral panic. In Seigel, M. (Ed.), *Panic, transnational cultural studies, and the affective contours of power* (pp. 27–44). Oxfordshire: Routledge.

Dua, J. (2019). *Captured at sea: Piracy and protection in the Indian Ocean* (Vol. 3). Oakland: University of California Press.

Duncombe, L. S. (2017). *Pirate women: The princesses, prostitutes, and privateers who ruled the Seven Seas*. Chicago: Chicago Review Press.

Duncombe, L. S. (2019). *A pirate's life for she: Swashbuckling women through the ages*. Chicago: Chicago Review Press.

Gilmer, B. (2014). *Political geographies of piracy: Constructing threats and containing bodies in Somalia*. Berlin: Springer.

Gilmer, B. (2017). Hedonists and husbands: Piracy narratives, gender demands, and local political economic realities in Somalia. *Third World Quarterly, 38*(6): 1366–1380.

Hayward, K. (2004). *City limits: Crime, consumer culture and the urban experience*. London: GlassHouse.

Hesse, B. J. (Ed.). (2013). *Somalia: State collapse, terrorism and piracy*. Oxfordshire: Routledge.

Hunter, M. (2002). Rethinking epistemology, methodology, and racism: Or, is White sociology really dead? *Race and Society, 5*(2), 119–138.

Hurlburt, K. (2013). The human cost of Somali piracy. In M. Mejia & J. Schroder-Hinrichs (Eds.), *Piracy at sea* (pp. 289–310). Berlin, Heidelberg: Springer.

Irwin, K., & Chesney-Lind, M. (2008). Girls' violence: Beyond dangerous masculinity. *Sociology Compass, 2*(3), 837–855.

Klausemann, U., Meinzerin, M., & Kuhn, G. (1997). *Women pirates and the politics of the Jolly Roger*. London: Black Rose Books.

Lambert, D., Martins, L., & Ogborn, M. (2006). Currents, visions and voyages: Historical geographies of the sea. *Journal of Historical Geography, 32*(3), 479–493.

Leeson, P. T. (2009). *The invisible hook*. Princeton, NJ: Princeton University Press.

Marchione, E., & Johnson, S. D. (2013). Spatial, temporal and spatio-temporal patterns of maritime piracy. *Journal of Research in Crime and Delinquency, 50*(4), 504–524.

Maton, K. (2013). *Knowledge and knowers: Towards a realist sociology of education*. Oxfordshire: Routledge.

Mohanty, C. (1988). Under Western eyes: Feminist scholarship and colonial discourses. *Feminist Review, 30*(1), 61–88.

Moore, R., & Muller, J. (1999). The discourse of 'voice' and the problem of knowledge and identity in the sociology of education. *British Journal of Sociology of Education, 20*(2), 189–206.

Nyman, E. (2019). Techno-optimism and ocean governance: New trends in maritime monitoring. *Marine Policy, 99*, 30–33.

Ong-Webb, G. G. (2006). Piracy in maritime Asia: Current trends. In P. Lehr (Ed.), *Violence at sea* (pp. 49–106). Oxfordshire: Routledge.

Peters, K. (2010). Future promises for contemporary social and cultural geographies of the sea. *Geography Compass, 4*(9), 1260–1272.

Peters, K. (2016). *Water worlds: Human geographies of the ocean*. Oxfordshire: Routledge.

Smart, C. (1995). Feminist approaches to criminology, or postmodern woman meets atavistic man. In C. Smart (Ed.), *Law, crime and sexuality* (pp. 32–48). London: Sage.

Smith, D. E. (1974). Women's perspective as a radical critique of sociology. *Sociological Inquiry*, *44*(1), 21–33.

Snider, L. (2003). Constituting the punishable woman: Atavistic man incarcerates postmodern woman. *British Journal of Criminology*, *43*, 354–378.

Stanley, J., Chambers, A., Murray, D. H., & Wheelwright, J. (Eds.). (1995). *Bold in her breeches: Women pirates across the ages*. Leixlip: Rivers Oram Press.

Steinberg, P. E. (2001). *The social construction of the ocean* (Vol. 78). Cambridge: Cambridge University Press.

Townsley, M., Leclerc, B., & Tatham, P. H. (2016). How super controllers prevent crimes: Learning from modern maritime piracy. *British Journal of Criminology*, *56*(3), 537–557.

Turley, H. (1999). *Rum, sodomy, and the lash: Piracy, sexuality, and masculine identity*. New York: NYU Press.

Twyman-Ghoshal, A. A. (2014). Contemporary piracy research in criminology: A review essay with directions for future research. *International Journal of Comparative and Applied Criminal Justice*, *38*(3), 281–303.

United Nations. (1982/2022). Part VII, High seas, Article 101. Piracy. United Nations Convention on the Laws of the Sea. https://www.un.org/depts/los/convention_agreements/texts/unclos/part7.htm.

United Nations. (2004). United Nations convention against transnational organized crime and the protocols thereto. https://www.unodc.org/documents/treaties/UNTOC/Publications/TOC%20ConventionTOCebook-e.pdf.

Vagg, J. (1995). Rough seas? Contemporary piracy in South East Asia. *British Journal of Criminology*, *35*(1), 63–80.

Young, J. (2007). *The vertigo of late modernity*. Los Angeles, CA: Sage.

1 Gender and the Female Pirate Imaginary

It has never been easy to be a woman. Historically, those who controlled maritime commerce were hostile to women's labor and actively discouraged or forbade women from working aboard ships because of the misguided belief that women could not handle seafaring's physical demands and precarious lifestyle. Never mind that the poor and working-class women who might choose to work aboard a ship faced far more dangerous and precarious situations in the labor available to them on shore, such as selling sex, cleaning and caring for richer people's homes and families, or market trade, all of which subjected them to sexual harassment, social opprobrium, abysmal living conditions, and a short life expectancy.

The male colleagues of the extraordinarily brave poor and working-class women who chose shipboard work routinely subjected the women to cramped, unhygienic spaces that lacked privacy, bullying and intimidation by the majority male crews (Appleby, 2013). Having excluded women from the opportunity to earn a living aboard a ship, these seafaring men were free to celebrate their labor as a masculine feat and, in so doing, further reinforce gendered assumptions of a seafaring "brotherhood" (Garber, 2021). Beyond promoting inaccurate gender stereotypes that depicted women as unable to endure the trials and tribulations of the sea, sailors promoted myths that women's very presence on ships brought bad luck: storms, mechanical failures, and even infighting among crew members (Appleby, 2013). Whereas society has continuously told women they do not belong at sea, one cannot help but note the irony underpinning the profound feminization of the sea and the maritime vessels, typically referred to her as *she* and *her* and treated as womblike vessels for the social and material transportation of men across the maritime space (Ciobanu, 2006).

Despite their underrepresentation at sea, women have a long-standing work association with seafaring lifestyles and industry (Parker, 2009). Most contributed their labor in activities based along the shore or directly linked to their seafaring spouses and families, while some cultures such

DOI: 10.4324/9781003225201-2

as marginalized groups along China's southern coast engaged women to actively engage in fishing, shipping, maritime smuggling, and piracy (Kwan, 2020). Communities like the latter did not envision the sea as a gendered space, and therefore have a rich history of women's involvement in maritime activities. Advancements brought by the industrial era also contributed to a slow increase in the number of women seeking employment at sea. Women continue to work in ways that benefit maritime industry by providing provisions and retail for seafarers, assisting with family- or community-owned fishing and oyster companies, and laboring along the waterside in a wide range of capacities. However, cultural taboos, the prevailing gendered social organization of labor, and enduring prevalence of unequal relationships that disproportionately burden women with caregiving responsibilities continue to dissuade women from going to sea.

Gendered assumptions about the sea impact how maritime spaces, actors, and activities are studied and imagined. If the sea is conceived of as a man's space, then any woman within that space is constructed as problematic because she is out of place. Her intrusion into that space is rationalized as unintentional—she was pushed into the circumstance due to extreme poverty or social obligations to support her family, or her presence becomes further evidence of her defiant nature and inability to conform to social norms. The woman pirate imaginary—the fanciful image conjured up when one thinks of what a woman pirate looks like, how she acts, what she does, that is, both a product of and reinforced through literature, film, and even piracy studies—amplifies this false dichotomy through two seemingly opposing constructs. She is either normalized as a "victim" without agency or a fantastical "leader/fighter" who exercises too much agency in her willful disregard for others. This chapter explores the gendering of piracy as both a lived reality and a political project that reproduces popular imaginaries of how women experience and perform the violence of piracy.

Victims

Debbie and her partner Bruno were now becoming sick on a regular basis. The holy month of Ramadan had arrived, and their captors were fasting between sunrise and sunset. The heat was becoming unbearable and the leftover food they were fed was mostly spoiled. A pile of cockroach carcasses was forming in the corner of their room from their daily sweepings. They had already nicknamed this house "the cockroach house". Bruno had picked an empty plastic bag from a thorn bush outside while they were washing their clothes and was now attempting to sweep the pile into it when the men came barging into the room. Debbie saw the scraps of clothing in one of their hands and knew they were about to be blindfolded again. They shoved Bruno against the wall, violently pressing his face against the dusty mortar bricks. They tied his hands behind his back instead, as if they wanted him to see what was about to unfold. Debbie saw the look of panic on his face as they led her out of the room.

Terrified, Debbie was led across the small courtyard and shoved through the doorway of an empty room. A single mat was laid on the floor in the far corner. "Sit down. You must not speak", her captor shouts in her face. "Do you understand? You must not speak". Debbie nodded in agreement. A few seconds later another man entered the room and directed the first man to leave. As he hurried out, Debbie felt an overwhelming sense of dread. She was now alone with the pirate who had been harassing her for weeks. He approached her, swung his arm, and landed a blow to her head that knocked her sideways onto the floor. He climbed on top of her and began pinching her legs and breasts while continuing to punch her. Debbie used her arms to protect her head, hoping he would tire of the beating. As he began to strangle her and try to force her legs open, she knew her worst fears were coming true. She was going to be raped. The harder she fought, the more he tightened his grip. Panicking because she could no longer breathe, Debbie allowed him to open her legs and attempted to block out what was happening to her.

In 2010, Debbie Calitz and her partner were kidnapped by Somali pirates while sailing in the Indian Ocean. They were taken to Somalia where they were held captive and moved around a network of pirate hideouts for 20 months. They lived in dark rooms, were beaten, and in Debbie's case, raped. They were eventually released and recovered after the Italian Intelligence Service (AISE) presumably paid their ransom, and Debbie would go on to write an autobiography detailing her experiences as a piracy hostage (see Calitz and Hill, 2013; MacAskill et al., 2015). Her story is somewhat unique to contemporary piracy, as most contemporary piracy hostages, just like most pirates, are male seafarers from the Global South. However, her testimony is an important part of understanding women's experiences of victimization in relation to piracy because it emphasizes how those experiences have been politicized and instrumentalized throughout history. Even Debbie's story is told through his lens of a religious reawakening spurred by surviving the horrors committed by Third World criminals. Stories of fragile, passive females victimized by pirates are often juxtaposed with the hypermasculinized typology of their abuser—the rogue villain, heavily armed and lusting after wealth and pleasure—to garner public support for bringing pirates to justice. Conflating women with victimhood for practical and entertainment reasons is not unique to maritime piracy, and feminist criminologists critique the discipline of criminology for continuously portraying women as vulnerable victims while ignoring the possibilities of women as active agents of crime (Barberet, 2014).

Women's experiences with victimization related to piracy vary based on the time period, the geographic location, the type of piracy, and the identity of the victims. Historical records reveal women experienced different degrees of physical, verbal, psychological, and emotional abuse that often took the form of domestic slavery, concubinage, and sexual servitude (Appleby, 2013). The following section explores women as victims of piracy by engaging with three broad themes consistent across time periods and locations of piracy: (1) violence against women as normalized by pirate toxic masculinity, (2) women's bodies as sites of broader struggles, and (3) varied experiences of victimhood based on identity. A commonality among these themes is that female captives were not *passive* victims of piracy-related violence (Bekkaoui, 2011). Rather, they resisted, fought back, and drew upon their gendered identities as women, wives and mothers to assuage the violence, attempt escape, or earn their freedom when possible (Tucker, 2014).

Contextualizing Piracy's Normalization of Violence against Women

The violence of piracy must be understood within the broader sociohistorical period in which it took place. Documents detailing the nuances of piracy beginning in the 1500–1700s and during the so-called "Golden Age" of piracy in the 1600–1700s when piracy was flourishing in the Caribbean describe a context of increasing maritime disorder and endemic societal violence coterminous with expansion, conquest, and the development of states (Appleby, 2007). The violence of piracy, then, was an extension of broader social violence where, for example, public executions and other forms of punishments were the cornerstone of judicial institutions (Ruff, 2001). Broader cultural codes of masculinity and femininity were also impacted by this violence. In the patriarchal gender logic of the time, intimate partner violence, domestic abuse, and rape were integrated into relationships between spouses and sexes that reinforced an acceptance for men holding power over women and a man controlling his wife (Lidman & Malkki, 2018).

Pirates' gendered acts of violence aboard ships and within coastal communities reflected and reinforced prevailing cultural logics of masculinity and femininity (Tucker, 2014). The coastal and shipboard culture created by pirates celebrated (some forms) of violence against men who were wealthier, seen as representatives of unjust governments, or who were identified by their crew as particularly cruel. Other, less common pirate groups practiced indiscriminate violence against anyone who crossed paths. Violence against women, however, was routine but generally more hidden because it was less likely to take place in the presence of others.

Pirates used violence and threats of violence as mechanisms of control by subjecting male captives to violence and threats aimed at dishonoring them into submission and threatened women with the possibilities of sexual violence. Yet pirates' sexual assault of female captives, like piracy itself, is the product of the meanings its sociohistorical context associates with sexuality and sexual violence. For example, in the case of early modern Mediterranean piracy, Muslim Barbary pirates did not sexually assault Christian female captives if there was a possibility she would convert to Islam and become a wife (McDougal, 2005). Many women captives recognized this as an opportunity to spare themselves from abuses and agreed to convert and marry. These women demonstrated that being a victim does not mean being devoid of agency. Rather, they actively drew upon gendered perceptions of vulnerable femininity and identities as potential wives and mothers to navigate the threat of violence at sea and onshore with pirates.

Historical retellings of piracy often use descriptions of violence against women as defining attributes of male pirates; consider, for example, the frequent association of the troubling phrase "rape and pillage" with these accounts. Stories of the treatment of female captives thus become reference points for historians to understand the leadership and operational styles of pirate leaders and groups. For example, Frohock (2018) compares stories of sexual violence against female captives in Caribbean piracy during the 1700s. The English privateer Captain Woodes Rogers, who was a slave trader and Governor of the Bahamas, describes his crew's encounter with a group of Spanish women while searching for valuables as highly eroticized—the women's hair is neatly dressed, and their bodies thinly veiled with silks. This erotically charged description underscores the ostensible restraint and civility demonstrated by Captain Rodgers and his crew as opposed to the popular understanding of pirates running wild among coastal communities and raping their women. In Captain Rodger's account, the fact that his crew did not sexually assault women was somehow proof of their personal merit.

In a marked contrast to privateers, buccaneers had a reputation for brutal sexual abuse and flagrant homosexual behavior during a time when most societies prohibited it. One buccaneer, Alexandre Exquemelin, was known for his violent aggression toward women. He and his crew forced poor, working-class locals to give their women to them as concubines or purchase them for prostitution aboard their ships. Similarly, *A General History of the Pyrates* depicts the brutal treatment of non-European women at the hands of pirates (Dafoe, 1999). The pirates would live among the local populations for several weeks at a time committing "outrageous Acts" against the women and eventually killing most of the locals and setting the town on fire

(Dafoe, 1999: 117). The pirates open brutalization of women signals a "systematic victimization of female captives" that runs counter to many pirate codes that prohibit the abuse of women (Frohock, 2018: 138). Such codes, which were aimed at establishing a more utopian, maritime counterculture to the bourgeois-controlled hierarchy on land, while agreed upon prior to setting sail, were not always realized in practice as women's bodies became a site in which broader struggles of power played out.

Women's Bodies as Sites of Struggle over Power, Society, and Culture

Violence is a profoundly embodied experience. Feminist political scientist and international relations scholars have long argued for bodies to be taken seriously as sites of political struggle (Ahall, 2012; Wilcox, 2015). Modes of violence can construct, use, and target bodies in ways that strategically shape power dynamics, culture, and political terrain. Violence against women in maritime piracy has also been utilized to shift power dynamics not only at the individual level among crewmembers but also at broader scales between ships and nation states.

Arguably, one of the most well-known examples of the politicization of women victimized by pirates is the hijacking and rape of women passengers aboard the Mughal emperor Aurangzeb's trading vessel, the *Ganj-i-Sawai*. At the end of the 1600s, an English pirate group led by Henry Avery boarded and ransacked the ship and reportedly tortured the men and raped the women, including the emperor's daughter. This incident, particularly the rapes and defilement of the women, created an international crisis that jeopardized the tenuous trading relationships between the British East India Company and the Mughal Empire. The emperor demanded the British government take responsibility for the actions of the English-born pirates or risk losing exclusive rights of trade in the Indian Ocean (Dua, 2018). The abuses against the women became the focal point of the highly publicized trial of Avery and his men and obtaining justice for the women became synonymous with repairing diplomatic relations between England and the emperor.

There are other examples of pirates utilizing indiscriminate acts of violence toward women and men as a strategy for garnering fear, respect, and thus, consolidating power. Perhaps, most notorious for this was Edward Teach, commonly known as Blackbeard. Teach boasted about his unpredictable aggression toward the coastal communities of the American Carolinas (Defoe, 1999: 139). On some occasions he treated the inhabitants well and traded with them, but on other occasions he brutally raped their daughters and wives. Teach utilized unpredictable violence to keep those around him on edge and as a means for demonstrating that he alone held all the power. His unpredictably made him an ever-present threat and earned him the reputation of someone not to be trifled with (Frohock, 2018).

The extent and reach of Teach's sexual brutalization of women included not only the unsuspecting strangers of coastal communities but also his own wife. He describes raping his new bride on their wedding night and then forcing her to have sex with several of his crewmembers. Teach sat and watched as his new wife was repeatedly raped by the other men. Despite Teach claiming this custom of rape as an example of his contempt for social expectations, other scholars have theorized his pleasure in watching the rapes as evidence of homoerotic desires and as a demonstration of self-empowerment by choosing to become a cuckold (Frohock, 2018). The woman's body becomes central to this power struggle over homoerotic intentions and self-aggrandizing. One could also argue that the pirates' gang rape is less about them having homoerotic desires than it is about the tight group bond they need to maintain to stay loyal to each other and keep committing crimes together; in a symbolic way, gang rape makes them of one unified body through sex with a single woman. This form of group violence toward women is also demonstrated through the more contemporary example of biker gang rapes of women (Quinn, 1987).

Geopolitical contestations have also been fought on women's bodies. At the beginning of the nineteenth century, northern Europeans attempted to distance themselves from their prior participation in piracy by associating piracy with men from other cultures, namely brown-skinned men from North Africa. Underscoring the violence committed by these brown men against white women was critical to framing pirates as uncivil and barbaric. Like other civilizing projects of this period, Western Europeans' depictions of race, religion, and civic backwardness enabled their states to embark on civilizing missions throughout Africa and Southeast Asian regions (Moberly, 2011). The gendered violence of piracy became a tool to provoke and justify battles. The white, protestant men of Europe were obligated to prove their masculinity by saving their women from the brown, Muslim, Barbary pirates of North African shoreline. As argued by Moberly (2011), "pirate violence against women, in its final Mediterranean chapter, was reconfigured to serve the interests of European conquest of the southern shores" (34–35). Thus, the construction and delivering of stories of pirate violence against women helped legitimize equally violent efforts aimed at developing, preserving, and expanding the modern state.

Women's bodies also became sites of cultural struggle where group identities were challenged or reinforced. Dominant culture during this era typically depicted pirates as counterculture figures who carefully crafted their very existence and lifestyle in opposition to the land-based society.

However, the libertarian ethos of unregulated commerce and freedom so central to these dominant cultural depictions of pirates is considerably challenged by stories of the brutalization of women on ships and in communities on shore. Leeson (2009) argues that pirates advocated for equality, freedom, and independence only so far as it was economically advantageous to the male members of their group. Similarly, Frohock (2018) contends that pirates followed pirate code when it was individually advantageous to do so, as evidenced by pirate groups amending their codes to condone sexual assault of so-called "fallen" women—unmarried, nonvirginal women. This could mean women who were sex workers or women whose husbands had died or abandoned them. Permitting sexual abuse of some—but not other—women's bodies reinforces cultural gendered expectations of female virginity as a moral virtue and, in this case, a protective factor—a topic discussed in greater depth in Chapter 3.

Varied Victimhood: Nation, Race, and Purity

Most studies of women's interactions with pirates portray them in generalized categories that reduce women to their reproductive and caregiving functions such as prostitutes, slaves, captives, or servants (Frohock, 2018). Yet the women's embodied racial, national, and class identities cut across these categories to impact the scope and type of violence they endured from pirates. Early modern captivity literature depicting women's experiences in captivity was widely written by male European writers for readers in Europe (Moberly, 2011). It is not surprising, then, that these European accounts of Christian women held captive by Muslim Barbary pirates and pirates of the Ottoman Empire depicted the women as sexually deviant traitors who betrayed their religion and nation rather than as savvy strategists who seized the opportunity to secure their survival by marrying their captors. Narratives of these "traitors" were accepted and unchallenged by an English population steeped in patriarchal understandings of women as unstable and fragile, thereby reinforcing portrayals of the captives as weak, hopeless women.

Eighteenth-century Dutch women held captive by Maltese pirates and Moroccan corsairs, privateers of the Barbary Coast, also challenged notions of women as frail, passive victims. When confronted with violence including physical and sexual assault or death, these women resisted and requested death or being ransomed to other captors rather than marrying the sultan (Tuckers, 2014). The women chose defiance. There are also examples of women from this period who chose execution alongside their captured husbands rather than be taken captive, and one woman even managed to negotiate the release of both her and her husband by agreeing to convert to Islam. Women confronted and navigated pirates' violence in ways that would be viewed noble and patriotic had they been men. As women, however, they received no such accolades.

Race also plays a central role in shaping the realties and perceptions of women's victimization. The victimization of white European, women by pirates has continuously been framed as a most heinous crime committed not only against the women but also against society. The victimization of women of color, however, is depicted as less problematic. Exquemelin infamously ranked ethnicities in a hierarchy and compared non-European women to animals, claiming they were less than fully human. Although the pirates' perceptions of racially based characteristics varied by pirate group and historical period, those perceptions influenced the interpersonal interactions between the pirates and women and whether they were treated with appreciation and respect or met with disregard and abuse. For example, Frohock (2018) underscores the conflicting tales of European pirates' perceptions of African women as devoted companions because of their supposed willingness to travel on voyages with and warn the pirates of danger, which could be read as a white savior narrative, or one of contempt and abuse involving the rape and pillage of native populations.

Afro-Caribbean women's experiences of being held captive by pirates provide unique insights into how women freed themselves from captivity through marriage and motherhood. Parish records and other historical documents provide a glimpse into the agency of Afro-Mexican women who were kidnapped from Saint-Domingue (Haiti) by buccaneers in the late 1600s. These women were purportedly kidnapped to provide a multitude of gendered and racialized services including sex and domestic labor, along with motherhood and companionship (Sierra, 2020: 3). While some might argue that the women essentially traded one form of captivity for another by living with the buccaneers and French colonists. Within this new realm of captivity as wives, though, the women survived by marrying the European men and giving birth to mixed race children—ultimately earning at least some form of freedom as mothers.

Society has long treated women's sexual histories as a measure of their worthiness for the status of an "ideal victim" worthy of justice (Christie, 1986; Bosma et al., 2018). Prizing women's virginity and sexual fidelity is common in societies that embrace patriarchal ideas that have persisted throughout history. As early as medieval times, perpetrators of rape and sexual violence against women who were either married or assumed to be virgins faced severe criminal sanctions. These same socio-legal norms, however, regarded rape as an impossibility for women who were unmarried and/or believed to have multiple sexual partners. Both principle and practice made it clear that legal protections were conditional upon a woman's sexual experience and marital status (Lidman & Malkki, 2018).

The treatment of women aboard pirate ships in the 1700s in the Mediterranean reflects the deeply rooted sexual dichotomy of unchaste/pure women that was pervasive in society at the time. Articles agreed upon and drawn up to guide behavior and settle disputes aboard the pirate ship *Revenge* described, "If at any Time we meet with a prudent Woman, that Man that offers to meddle with her, without her Consent, shall suffer present Death" (Defoe, 1999: 343). Thus, the crew set out to sea in agreement that "prudent" women were a protected class. Sexually assaulting or raping a prudent woman would be met with capital punishment. Yet, the rape and sexual assault of unmarried women believed to have multiple sexual partners generally was not even seen as a crime. As Frock contends, "from a piratical viewpoint, women who transgress sexual rules of propriety forfeit their rights to their bodies and enter a lawless realm that affords them not new liberties and empowerment but rather new ways of being victimized with impunity" (2018: 137). The differentiation between unchaste and pure women enabled the violent victimization of female captives despite rules put in place to protect against such actions. Although it is important to note that incidents of pirates raping women and throwing their bodies overboard challenge claims of pirates' libertarian approach to equality and self-rule and suggest that being a married woman, or a woman believed to be a virgin, was not always a protective factor as pirates like Edward Teach were known to show no mercy on their victims (Dafoe, 1999).

The victimization of women (and men) by pirates is horrific and deserves attention and action. However, it is important that women are not only viewed as *victims* of piracy. Women may also be willing, active participants in piracy who victimize others—including both men and women, as I discuss at length in Chapter 4. Despite the myriad of ways that women contribute to piracy's operational success, the role of women pirates as leaders/fighters remains the most romanticized and profligate. Women leading pirate fleets and fighting alongside men at sea are a historical reality that will be examined further in the next chapter. To understand the complexity of this historical reality, the remainder of this chapter focuses on what I call the *female pirate imaginary*, a conceptualization of a female pirate based on real historical figures repackaged and exotified for (male) public consumption.

The Female Pirate Imaginary

The coastal breeze was doing nothing to provide reprieve as the scent of decay hung in the dry, Somali air. Locals generally avoid visits to the fish market and purchase fish from the more central markets at a higher price because they say it takes days to get the smell out of your clothes. Men are busy cleaning their fish on makeshift tables set up in the sand, hoping to make a profit by the day's end. They smile generously and point in a direction further up the beach. There, in the sand next to a stream of maggots making their way to the water, lies the catch of the day. The massive fish is half covered in sand just centimeters from a pile of smaller rotting fish corpses. As the Port of Bosaso stirs with fisherfolk untangling fishing nets, the sounds of dull knives striking against flesh and wood and the buzz of informal commerce add a sense of excitement to the moment. One's eyes cannot help but be drawn to the bright colors and beautiful patterns of garbasaars (traditional shawls) adorned by Somali women as they move about in the shallow waters.

Khadra uses one hand to help move her garbasaar, now fully saturated, along in the water as she shifts the net full of fish from her side to her back. The sandy muck along the sea's floor seeps over the top of her foot and grabs hold of her harder with each step. Her legs tire as she trudges slowly to the shore, but she is thankful there are fish today. She's been carrying fish into shore for the fishermen for over 30 years and with each year, the fish stocks are decreasing. Her husband, a fisherman and the reason she began transporting fish, said it's because of all the illegal foreign fishing trawlers from Europe and Asia who come and steal their fish in the dark of the night.

Somali women have long carried out the important task of transporting fish from boats to shore. When I asked why women fill this role, a local fisherman responded that they always have, noting the vague "it is tradition" that ethnographers all over the world hear after asking a question that seems obvious to local people. Women venturing out to sea on the boats, however, have always been viewed as a man's task that is too dangerous for women. The Somali women either giggled at the prospect of going to sea or gasped in horror at the absurdity of the idea when I brought it up. Instead, they are proud to continue contributing to the fishing community by serving as the link between boats and shore. Understanding women's involvement in contemporary piracy off the coast of Somalia begins to make more sense when it is viewed within the broader context of women's boundary work—particularly among Somalia's coastal and fishing communities. It is impossible to not see these women as important links between the communities, fisherfolk, and pirates. Yet, images of these women rarely come to mind for most people when discussing piracy. Why not? What imaginaries emerge instead?

The term "pirate" often conjures up images of bearded, burly white men wearing puff sleeved shirts and battle-torn trousers while brandishing a gun or a sword. They stand perched at the bow of a wooden vessel with a menacing black jolly roger flag waving in the background, sometimes with a loyal tropical bird perched on a shoulder. Often these imaginary pirates have significant disabilities, such as a missing leg clumsily substituted with a wooden prosthetic or a damaged eye covered with a ragged cloth patch. In popular culture, these disabilities arguably aid in depicting these fictional pirates simultaneously as men who ruthlessly face danger while rejecting regard for their own or others' safety as well as dominant social norms regarding propriety. No equivalent set of images exist for women, who seldom featured in this popular imaginary despite the reality that female pirates have captured the attention of readers and viewers throughout history. Narrative and illustrative depictions of female pirates and their shipboard roles began as early as the 1700s and has slowly evolved over time. Over the next three centuries, the female pirate imaginary has both reflected and coproduced broader social understandings and cultural expectations of gender and sexed bodies through the lens of maritime crime.

In the early eighteenth century, British publishing houses were instrumental in producing and distributing books featuring pirates as their central characters. The bestselling monograph, Daniel Defoe's (1724) *A General History of the Robberies and Murders of the Most Notorious Pyrates*, was hugely influential in producing a pirate imaginary for public consumption that would ultimately frame the way Western Europeans and North American public thought about pirates for the next three hundred years. It pandered to British society's desire for the exotic, grotesque, and violent with stories of infamous pirates pillaging communities and brutally raping and murdering captives. This violence was woven into broader themes of brotherhood and liberation that framed pirates as antiheroes to be both celebrated and feared (Campbell, 2016). The demand for antiheroes and violence also gave rise to the sensational, cheap, and heavily illustrated "penny bloods" (also known as "penny dreadfuls") booklets. The Victorian public went wild for their gothic tales and grisly true crime that included pirate adventures because they provided escapism at an affordable price while simultaneously denouncing rigid class hierarchies and celebrating poor and working-class people getting over on richer folks.

The woman pirate imaginary also emerged during this time and slowly underwent a noticeable visual and behavioral transformation as understandings of sex and gender changed in the Western societies where the literature originated. These changes materialized in the way the female pirate and her body were presented and discussed in both narrative and illustrative forms. Two female pirates, Anne Bonny and Mary Read, were introduced to a wide audience of readers via *A General History* (Dafoe, 1724) and provide a starting point from which to explore these changes overtime. Historical accounts of Bonny and Read are based on trial transcripts, government proclamations, and other official reports. The authenticity of these women's existence is not debated, but the accuracy of the details in the recounting of their stories may contain fictive elements.

Accounts of Mary Read's life usually begin by attributing her early maritime experience aboard a Man of War, a powerful warship, at the age of 13 to her ability to disguise herself as a boy with sailing skills, which she used when a ship she was aboard was taken by pirates. She later joined the pirates, still disguising her gender, and earned a reputation as a resolute, brave, and fierce fighter. It was another forward and courageous woman pirate, Anne Bonny, who eventually discovered the truth about Read's identity. The two developed a strong bond and were able to keep Read's gender a secret until one day Read accidently showed her "very white" breast to a male companion. Literary scholars point to the revealing of Read's breast in text and illustration as both a symbol for and a harbinger of the growing politicization of women's bodies and the separation of gender performance from sexed bodies taking place in Victorian England (O'Driscoll, 2012; Perry, 1991).

Read and Bonny put on the same clothes and bravely stepped into the same roles as male pirates to steal and terrify for a living. Illustrations of Bonny and Read in *A General History* show the women dressed in full pants and long-sleeved coats, each armed with swords and axes. Nothing draws attention to their female bodies nor problematizes their agency as pirates. Rather, their womanhood is mobilized through narrative accounts of their inability to resist their passions for male counterparts. Each woman eventually falls in love with and marries a pirate, proving—at least to conventional readers of the time—that despite costuming their gender to pass as men, they are unable to overcome their underlying desire to participate in heterosexual love. Society's expectations remain intact.

The 1725 Dutch reprint of *A General History* contained new illustrations of Read and Bonny that put their gendered bodies unavoidably on display. Read and Mary feature prominently on the title page of the book, perched atop a ship deck with a piece of cloth wrapped around the lower halves of their bodies while their breasts are fully bared. They wield their swords as a black pirate flag flutters in the wind behind them. Inside the book, the chapters about Read and Bonny now feature individual illustrations of each woman. Read stands atop a Jamaican cliff dressed in trousers, her shirt unbuttoned, and breasts revealed as she holds a sword in one hand and presses her hat against her head with the other. Bonny is also depicted atop a Jamaican cliff wearing trousers and a breast-baring open blouse as she fires a gun in one hand while brandishing a sword in the other.

During this era, literature increasingly desexualized depictions of "domestic" women in while eroticizing women who broke the law. Thus, we can locate the beginnings of the normalization of scopophilia toward and sexualization of women pirates as far back as the eighteenth century. O'Driscoll (2012) brilliantly argues that the presentation of criminal women in late seventeenth- and eighteenth-century literature encourages readers to view them as abject bodies and objects of public scrutiny. Unlike their feared yet respected male counterparts, women pirates are reduced to their gendered bodies and their breasts become sites of cultural speculation for public consumption. Despite the women's best attempts at passing as men, the women are betrayed by their breasts. Readers are reminded that the women pirates' agency is as fraudulent as their attempt at passing.

While European publishing houses began producing more exotic and erotic female pirates, across the ocean American literature presented a much more modest version of the female pirate imaginary. The female pirates of American pamphlet novels were gentle hearted and physically inferior heroines cloaked in fierce costume. The cross-dressing heroines were a product of the harsh new-world environment in which they were imagined. America had recently gained its independence and was amid realizing its own imperialist and expansionist efforts while also vying for recognition and respect in the international realm. Female pirates became symbols of America's immature, soft conquering approach that had not quite reached its glorious ambitions of expansion (conceived as manhood) (Anderson, 2016). Readers embraced these characters as they saw them as representative of their fledgling nation with the potential for greatness while still allowing for shortcomings due to immaturity and weakness, which dominant culture at the time equated with women's bodies.

In the twentieth and twenty-first centuries, representations of female pirates changed and expanded alongside shifting discourses of criminal women which were then packaged for male consumers and promoted widely via rapid technological advances. During the century prior, criminologists were invested in the study of criminals as born, not socially produced. Women offenders were portrayed as defective, products of inbreeding, and as biologically inferior to non-offending women (Goring, 1913; Thomas, 1923; Pollak, 1950). Women offenders, when compared to male offenders, were deemed more "deficient in moral sense" but stronger in "sexual instincts" (Lombroso & Ferrero, 1895). These "unadjusted girls" (Thomas, 1923) were understood as biologically programmed to behave irrationally which (male) criminologists contended was evidence that they were beholden to their bodies, sexuality, and reproductive roles (Snider, 2003).

Twentieth Century

The mid-twentieth century marked the birth of television and film as a central source of entertainment and (re)producer of popular culture. It also ushered in a female pirate imaginary that echoed aspects of what a criminal woman was understood to be in the decade prior—a morally deficient, sexually charged deviant. Now, however, rather than reading about her in text, the public could observe her on the big and small screen, scrutinizing her appearance, mannerisms, strengths, and moral and physical shortcomings. Broadly, the contemporary female pirate imaginary of mainstream entertainment that persists today is what I shall term a "lustful leader". She is no longer concerned about concealing her identity or being betrayed by a rogue breast. Rather, her gendered body is central to her identity. Her breasts are proudly displayed and barely contained thanks to a tightly cinched corset. A gun or knife is often holstered to her bare thigh—inviting the viewer's gaze while also warning of the dangers of seduction. A battle at sea becomes a tantalizing spectacle where a female pirate's revealing garments and bare skin are understood as a part of combat rather than a hazard.

The twentieth-century female pirate imaginary not only shed the desire for sexual ambiguity, but also embraced some of the gendered traits previously seen as shortcomings (e.g., erratic, emotional, unstable) or unfeminine (e.g., aggressiveness, violence, lustfulness). Whereas the embrace of these traits can be considered a nod to feminist progress, they are limited by the highly sexed bodies that signal a more extreme level of scopophilia than in previous centuries. Whether her gendered body is a reclaimed pronouncement of female agency or a commodified object for male consumption depends upon the specific pirate character and context.

Women pirates of twentieth-century films followed a similar "pirate chic" aesthetic popularized by male pirates in nineteenth-century Romantic literary traditions that portrayed male pirates as social outcasts and antiheroes dressed in dark trousers, white blouses, tricornered hats, and donning red sashes or bandanas to symbolize violence and foreshadow blood and death (Campbell, 2016). When women pirates made their appearance on film in the mid-1900s, their appearance did not stray far from the wildly popular (male) pirate imaginary. Examples of this are seen in *Anne of the Indies* (1951) and *Cutthroat Island* (1995), two of the most notable twentieth-century films featuring female pirates. The protagonist of *Anne of the Indies,* played by Jean Peters, wore loose fitting capri pants, brown leather calf-high boots, and a long-sleeved white blouse in her role as pirate captain. In several scenes her ears are adorned with gold hoops, and she sports a red bandana—a nod to the red sash worn by her male counterparts. She displays a strong command over her ship and is highly skilled with a sword. However, her maritime experience and fierce fighting skills prove to be no match for her emotions. She cannot help but fall in love with one of her captives, spares his life twice, and later sacrifices her own for his. Accordingly, the plot ultimately domesticates an otherwise independent character.

Anne's character does not demonstrate a desire to present gender ambiguity (central to female pirates of eighteenth- and nineteenth-century literature), but her clothing remains modest and reflective of that worn by her male counterparts. Her character is not overtly sexualized, but her carefully applied makeup, perfectly tousled long curly hair, and unrestrained emotions that set signal to viewers she is different than typical (male) pirates. Throughout the film, she wavers back and forth between rational, bold leader and stereotypical irrational female. She appears fierce and in control steering the vessel and in combat but cowers and become frazzled when she finds herself in the arms of her love interest. Her inability to control her emotions, particularly love and compassion, becomes her fatal flaw. Anne's character reflects broader social efforts aimed at preserving and reinforcing conservative gendered expectations against the rapidly changing society of 1950's America. As a female pirate, she symbolizes the independent, adventurous woman seeking to eschew domestic life. In the end, Anne sacrifices herself to save the captive and his wife (the sanctity of marriage) as they watch her violent demise from a distance—protecting the ideal woman and preserving domesticity.

The female pirate imaginary also emerged on the small screen during the late 1990s. The wildly popular television series *Hercules: The Legendary Journeys (1995–1999)* featured Nebula, played by Gina Torres, a female pirate captain who left her home and society because of the mistreatment of women. Nebula's feminist stance toward rule, which includes dismissing any male crew members who appear to support antiquated gender roles, was made more palatable through her appearance that reminded viewers that beyond her fierceness she remained a highly sexed object packaged for consumption by male viewers. She fiercely ruled the seas in minimal clothing wearing only a gold metal bra, tight leather pants, and gold jewelry. When Nebula becomes empress, she trades in her sultry sea attire for thing-high red fishnet tights, a red leather corset dress with matching gloves, and a red leather whip. Nebula presents arguably the most overtly sexualized and fetishized representation, signaling aspects of bondage, discipline, sadism, and masochism, of a female pirate to date. Her titillating continuously leaves viewers in suspense as to whether her seductive outfits will withstand each battle scene.

The women pirates of mid- and late 1990s film and television demonstrate the staying power of some of the popular imaginaries of pirates originating in previous centuries. Some of them wear similar attire as their male counterparts, but it is not done in a manner to hide their gender. Instead, the idea of a woman pirate captain becomes more normalized so long as the viewers are offered fleeting reminders that her body is not solely for battle, but it is also an erotic object for our viewing pleasure. Thus, the scopophilia that began centuries prior with the revealing of female pirates' breasts has expanded to include sultry, revealing clothing, dominatrix demeanors, and dangerous thighs. This expansion will continue into the twenty-first century with the release of Disney's blockbuster film series, *Pirates of the Caribbean*; the fantasy SyFy network miniseries, *Neverland*; and a reengagement with and reimaging of historical female pirates as seen in the Emmy Award-winning HBO series, *Black Sails.*

Twenty-First Century

The Pirates of the Caribbean series of five full-length feature films has monopolized the big screen pirate imaginary for the first two decades of the twenty-first century and is a testament to the public's continuous fascination with the mid-17th Century Carribean piracy. The film series is a hyperbolic reimagining of seventeenth- and eighteenth-century pirate life with a larger-than-life central character named Jack Sparrow, played by Johnny Depp. Sparrow quickly becomes a sex symbol and cultural icon that would have the public swooning with the release of each new film over the next 15 years. However, the men were not the only ones getting attention. The film series strategically introduced a diversity of female pirates that both reinforced and challenged the female pirate imaginaries of previous centuries.

The first film in the series, *Pirates of the Caribbean: The Curse of the Black Pearl* (2003), presents Elizabeth Swann, played by Keira Knightley, a beautiful, intelligent, independent-minded woman who undergoes a transformation from a refined young lady into a courageous pirate. Swann proves to be an innate leader and a quick study in seafaring and swordsmanship. She easily adapts to pirate life but must constantly fend off a barrage of men who seem to be unable to resist her beauty. Her clothes reflect her upper-class background, as she wears a modest layered black and gold embellished overcoat with pants that accentuate her female figure but keep her skin concealed. Although her wardrobe varies slightly between films, her seafaring outfits appear to be more conducive for sailing and fighting than for enticing men, suggesting that viewers should take her seriously as a pirate. On land, however, she sheds a bit of her fierceness and reminds us of her gendered female body by accentuating her supple bosom with carefully chosen dresses and well-placed tresses of curly hair. Swann was well received by fans and became a central and recurring character in three other films in the series.

Disney added three additional female pirates in their subsequent films that helped diversify the cast and storylines—Tia Dalma, a mystic Caribbean woman of color who practices a Hollywood version of voodoo; Mistress Ching, a blind elder Chinese woman whose character is a nod to the great historical female pirate Ching Shih; and Angelica Teach, a Hispanic woman purported to be Blackbeard's daughter. Dalma, played by Naomie Harris, first appeared in *Dead Man's Chest* (2006), and is arguably the most exotified of Disney's female pirates. She is portrayed as a flirty and playful woman with a thick West Indian accent. She speaks in riddles, exhibits quirky and capricious behaviors, and brings people back from the dead. Her skin glistens in her off-the-shoulder blouse, and her cleavage is accentuated by a carefully cinched corset. These aspects combine with her blackened teeth, tattooed face, and muted lace shawl to produce a mysteriously alluring yet creepy pirate imaginary that aligns with her backstory of being deceased but presently occupying an earthly body. This existential limbo materializes in Dalma's appearance that simultaneously attracts and repulses and leaves her occupying an uncomfortable space between sexualized and abject.

Mistress Ching, played by Takayo Fischer, is another female pirate who appears in the third film in the series, *At World's End* (2006). She is a ruthless, elderly woman who is blind yet successfully commands an enormous fleet. Chang's character is loosely based on the real-life female pirate, Ching Shih, who ruled her armada with an iron fist. Like Shih, Ching has strict rules for her crew that includes no deserting, no stealing from villagers, and no taking women as prisoners. Anyone who breaks these rules would be punished with beheading—the method of punishment also utilized by Shih. Ching has the stateliest appearance of all of Disney's female pirates. Her deep jewel-toned robe with gold embroidery envelops her body while turquoise jewelry adorns her neck, fingers, and eloquently wrapped hair. Her face is powdered white with bright red lipstick accentuating the center of her lips. Ching's stern facial expressions and serious demeanor command respect and suggest that years of piracy have taken a toll on her body and spirit. Uniquely, her character does not carry the same undertones of sexual objectification as the other female pirates. Rather, her age and poise suggest we take her seriously as a respectable pirate leader who has earned her place at the table among her male peers.

On Stranger Tides (2011) presents yet another complicated portrayal of female pirates in with Angelica Teach, played by Penélope Cruz. Teach is the daughter of the infamous pirate Blackbeard who has helped her develop into a seasoned pirate. Beyond her mastery of sailing and fighting, she is known for her trickery and deceit. We learn that she perfected her craft of lies, deception, and disguises from her former lover and the film series' protagonist, Captain Jack Sparrow. With Teach, we see a recycling of the feminine-yet-fierce female pirate combination popularized in twentieth-century literature. She wears skintight leather pants and knee-high boots coupled with an off-the-shoulder peasant blouse under a black corset that gives her a shapely appearance. When she is not wearing her black leather hat, her tousled hair is pinned back under a blue bandana. Teach's sultry, smoky-eyed take on pirate chic compliments her playful and naughty behavior that keeps Sparrow on his toes. Her staple gold cross necklace reminds us that despite her deception and violence, she is a woman of faith who demonstrates compassion when necessary.

Moving from the large screen to the small screen, two twenty-first-century television series featured women pirates in prominent roles—*Neverland* (2011) on SyFy network and *Black Sails* (2014) on HBO network. *Neverland* is a fantasy miniseries promoted as a spin-off of the popular children's book, *Peter Pan* (Barrie, 1928). Beyond the fantastical scenery and storyline, *Neverland* gives us the first mainstream female pirate villain, Captain Elizabeth Bonny. Bonny, likely meant to conjure images of the historical Anne Bonny, is an evil power-obsessed and lustful pirate who has menaced the seas for centuries with a goal of ruling Neverland for eternity. Her appearance is almost a mirror image of Disney's Angela Teach (both making their appearance in 2011). Her makeup perfectly captures the sultry smoky eye look, and she dresses in a white peasant blouse accentuated by a bosom enhancing corset, tight black pants, knee-high black boots, and alternating between a black hat or a red and tan sash atop her long, beautifully curled hair. Bonny's thirst for power and eternal life is her fatal flaw that leads to her disintegrating after encountering fairy mineral dust.

The history of Anne Bonny is again revived and repackaged three years later in the historical adventure television series, *Black Sails* (2014). The series attempts to be historically accurate in recreating the excitement and drama of mid-17th Century Caribbean piracy of piracy for mainstream viewing. Bonny is reimagined as a ruthless, cold-blooded pirate. She dresses in layers of clothes and wears a hat from which she glares out from at her adversaries. The intensity of her brilliant red hair is matched by her short temper and violent demeanor evidenced by her quickness to slit a man's throat or stab him to death. Like the early eighteenth-century accounts of female pirates, Bonny's downfall is when she becomes involved in a love triangle that triggers her uncontrollable jealousy and rage. This representation depicts her character flaw as a pirate as the result of her innate desires for a heterosexual relationship, which is in turn depicted as incompatible with piracy.

From Rogue Breasts to Dangerous Thighs

The evolution of the female pirate imaginary from the seventeenth century through contemporary times demonstrates how the bodies of (criminal) women become contested sites where society's expectations, insecurities, and fantasies materialize in appearance and behavior. For the most part, the female pirate imaginary popularized in literature, television, and film has been based on hyperbolic understandings of historical fiction loosely based on trial transcripts, government reports, screenplays, and so forth recorded and written by men for male consumption. One could argue that even the earliest (tamest) writings about Anne Bonny and Mary Read chose to recount personal and gendered details not equally recounted in stories of male pirates. Dafoe (1661/1992) discusses the women's passions and insecurities, the limitations of their sexed bodies, and the birth of their children. The topics, if they mentioned at all in tales of male pirates, are merely side comments rather than central to the disposition of their character.

The female pirate imaginary has consistently retained an inherently redemptive quality—both physically and emotionally. If she disguises herself as a man or wears modest clothing, an opportunity always presents itself for her to reveal her gendered body or reinvent her appearance in a more feminine manner. Similarly, her aggressive, violent, and brave actions are eventually quelled or overcome through her participation in heterosexual relationships or a decision to have mercy on captives. These physical and behavioral changes appear to take on a "corrective" quality that creates a space for the female pirate to once (re)occupy her presumptive *natural* gendered role that more closely aligns with social expectations.

Much like the male pirate imaginary, the female pirate imaginary cannot seem to shake the pirate aesthetic associated with the Golden Age of piracy. How female pirates are imagined and recreated on the big and small screens continues to be shaped by this time and geographical location. Even the most imaginative storylines will contain elements associated with the Golden Age of piracy because there continues to be a demand for it. The demand suggests this is how the public wants to imagine pirates, including female pirates, regardless of whether it reflects the realities of piracy on the high seas. Accordingly, the fetishization of female outlaws and criminals continues well into the twenty-first century as evidenced through the female pirate imaginary (Crane-Seeber, 2016). The women of piracy remain fantastical objects of desire that link pleasure and crime in erotic ways. The remaining chapters of this book attempt to reconstruct the realities of women in piracy so that we can begin to conceive of their agency and roles beyond that of victims and hypersexualized, scantily dressed leaders/fighters.

References

Åhäll, L. (2012). Motherhood, myth and gendered agency in political violence. *International Feminist Journal of Politics*, *14*(1), 103–120.

Appleby, J. C. (2007). The problem of piracy in Ireland, 1570–1630. In C. Jowitt (Ed.), *Pirates? The politics of plunder, 1550–1650*. London: Palgrave Macmillan. https://doi.org/10.1057/9780230627642_3

Appleby, J. C. (2013). *Women and English piracy, 1540–1720: Partners and victims of crime*. Rochester: Boydell & Brewer Ltd.

Barberet, R. (2014). *Women, crime and criminal justice: A global enquiry*. London: Routledge.

Barrie, J. M. (1928). *Peter Pan or the boy who would not grow up*. Kalamazoo: Kazoo Books LLC.

Bekkaoui, K. (2011). *White women captives in North Africa: Narratives of enslavement, 1735–1830*. London: Palgrave Macmillan.

Bosma, A., Mulder, E., & Pemberton, A. (2018). The ideal victim through other (s') eyes. *M. Duggan (red.), Revisiting the 'Ideal Victim'*, Bristol: Policy Press. 27–41.

Calitz, D., & Hill, U. (2013). *20 months in hostage hell*. New York: Penguin Global.

Campbell, M. (2016). Pirate chic: Tracing the aesthetics of literary piracy. In G. Moore (Ed.), *Pirates and mutineers of the nineteenth century* (pp. 25–36). Oxfordshire: Routledge.

Christie, N. (1986). The ideal victim. In F. Ezzat (Ed.), *From crime policy to victim policy* (pp. 17–30). London: Palgrave Macmillan.

Ciobanu, E. A. (2006). Early modern brave new world? *Analele Universităţii Ovidius din Constanţa. Seria Filologie*, *17*, 7–23.

Dua, J. (2018). Privateers and public ends: Piracy as global moral panic. In M. Seigel (Ed.), *Panic, transnational cultural studies, and the affective contours of power* (pp. 27–44). Oxfordshire: Routledge.

Frohock, R. (2018). Beyond bonny and read: Blackbeard's bride and other women in Caribbean piracy narratives. In N. Aljoe, B. Carey, & T. Krise (Eds.), *Literary histories of the early anglophone Caribbean*. New Caribbean Studies. Cham: Palgrave Macmillan. https://doi.org/10.1007/978-3-319-71592-6_7

Garber, L. (2021). Unsafe seas for women. In L. Garber (Ed.), *Novel approaches to lesbian history* (pp. 111–144). New York: Palgrave Macmillan.

Goring, C. (1913). *The English convict: A statistical study*. Martinez: HMS.

Kwan, C. N. (2020). In the business of piracy: Entrepreneurial women among Chinese pirates in the mid-nineteenth century. In J. Aston & C. Bishop (Eds.),

Female entrepreneurs in the long nineteenth century (pp. 195–218). Cham: Palgrave Macmillan.

Leeson, P. T. (2009). *The invisible hook.* Princeton, NJ: Princeton University Press.

Lidman, S., & Malkki, E. (2018). *Gender, violence and attitudes: Lessons from early modern Europe*. Oxfordshire: Routledge.

Lombroso, C., & Ferrero, G. (1895). *The female offender* (Vol. 1). D. Appleton.

MacAskill, E., Milne, S., & Swisher, C. (2015, October 8). Italian intelligence lied about hostage rescue to hide ransom payment. *The Guardian.* Retrieved from https://www.theguardian.com/world/2015/oct/08/italian-intelligence-lied-hostage-rescue-bruno-pelizzari-debbie-calitz.

McDougall, J. (2005). Savage wars? Codes of violence in Algiers, 1830s to 1990s. *Third World Quarterly*, *26*(1), 120.

Moberly, D. C. (2011). *Piracy, slavery, and assimilation: Women in early modern captivity literature* [Thesis]. Lincoln: University of Nebraska. https://digitalcommons.unl.edu/englishdiss/52/

O'Driscoll, S. (2012). The pirate's breasts: Criminal women and the meanings of the body. *Eighteenth Century*, *53*(3), 357–379.

Parker, M. (2009). Pirates, merchants and anarchists: Representations of international business. *Management and Organizational History*, *4*(2), 167–185.

Perry, R. (1991). Colonizing the breast: Sexuality and maternity in eighteenth-century England. *Journal of the History of Sexuality*, *2*(2), 204–234.

Pirates of the Caribbean: At world's end. Directed by Gore Verbinski. Walt Disney Pictures Jerry Bruckheimer Films, 2007.

Pirates of the Caribbean: Dead man's chest. Directed by Gore Verbinski. Walt Disney Pictures Jerry Bruckheimer Films, 2006.

Pirates of the Caribbean: Dead men tell no tales. Directed by Joachim Ronning and Espen Sandberg. Walt Disney Pictures Jerry Bruckheimer Films, 2017.

Pirates of the Caribbean: On stranger tides. Directed by Rob Marshall. Walt Disney Pictures Jerry Bruckheimer Films, 2011.

Pirates of the Caribbean: The curse of the Black pearl. Directed by Gore Verbinski. Walt Disney Pictures Jerry Bruckheimer Films, 2003.

Pollak, O. (1950). *The criminality of women.* Philadelphia: University of Pennsylvania Press.

Quinn, J. F. (1987). Sex roles and hedonism among members of "outlaw" motorcycle clubs. *Deviant Behavior*, *8*(1), 47–63.

Ruff, J. R. (2001). *Violence in early modern Europe 1500–1800*. Cambridge: Cambridge University Press.

Sierra Silva, P. M. (2020). Afro-Mexican women in Saint-Domingue: Piracy, captivity, and community in the 1680s and 1690s. *Hispanic American Historical Review*, *100*(1), 3–34.

Snider, L. (2003). Constituting the punishable woman: Atavistic man incarcerates postmodern woman. *British Journal of Criminology*, *43*(2), 354–378.

Thomas, W. I. (1923). *The unadjusted girl: With cases and standpoint for behavior analysis (No. 4)*. Boston, MA: Little, Brown.

Tucker, J. E. (2014). She would rather perish: Piracy and gendered violence in the Mediterranean. *Journal of Middle East Women's Studies, 10*(3), 8–39.

Wilcox, L. B. (2015). *Bodies of violence: Theorizing embodied subjects in international relations*. Oxford: Oxford University Press.

2 The Gendered Social Organization of Piracy

Although research on women offenders increased dramatically in the past half century, women as active participants in organized crime did not begin to receive attention until the early 1990s (Arsovska & Alum, 2014). Current research on women in transnational organized has been instrumental in exploring the central roles of women in sex trafficking and smuggling (Siegel & de Blank, 2010; Zhang et al., 2007), drug trafficking (Kleemans et al., 2014; Tregunna, 2014), and the mafia (Fiandaca, 2007; Allum and Marchi, 2018). Other than women who traffic women into the sex industry, women in organized crime typically make less money in less prestigious roles than men. Piracy is also a form of organized crime and yet, despite the enormous inroads made by these scholars, women's role in piracy, particularly as leaders and those committing acts of violence (e.g., "dirty jobs"), remains understudied.

This chapter attempts to strike a fine balance between deepening our understanding of female pirate leaders while attempting to avoid oversimplifying these women's complex lived realities. Our knowledge is limited by the small body of academic literature on the topic and the recognition that most of these women's stories and lives were recounted and historicized by men. Without autobiographical evidence, interviews, or testimonies, these women's legacies are not entirely their own, and at times, speculative at best. Yet, this chapter utilizes these histories to glean women's pathways to pirate leadership, highlight the similarities they share with women involved in other forms of transnational crime, and theorize whether women pirate leaders can be understood as maritime *edgeworkers*—women who voluntarily participate in high-risk activities as a form of escape and resistance in ways that challenge and reshape the boundaries of gender, society, and the sea.

DOI: 10.4324/9781003225201-3

Leaders and Fighters: Pathways to Power

Shortly after 5 am prayer in the town of Haradheere midway down the eastern coastline of Somalia, the streets are abuzz with men and women setting out about the daily business of surviving among the challenges of poverty, hunger, and now the violent radical Islamic group al-Shabaab. Hibaaq hurries home, her buibui's long black skirt grazing the tops of her sandals as she averts her gaze to avoid contact with the groups of men perched in their cheap plastic chairs chewing khat alongside the storefronts. She reaches her family compound and is greeted by a member of her household staff who informs her she missed a phone call. She sighs, slightly irritated and curious about what could be so important that it necessitates a call so soon after morning prayer. Stepping into the back room of her house, she sits down at the desk once occupied by her late husband. She recalls the extravagant lengths he took to import such a stately piece of craftsmanship from Dubai. She picks up the cellphone and sees two missed calls and a text from Abdirizak that reads, "send more ammunition".

Hibaaq is the first wife of the notoriously successful pirate leader-turned-investor, Abduwali, who recently passed away from tuberculosis. At the age of 32, she has now assumed his role as an investor who provides the initial funds and oversees the distribution of ongoing financial support to several piracy operations based in the area involving hundreds of people that will generate several million dollars in revenue. She picks up the phone and calls a trusted weapons and ammunitions dealer in Bosaso who tells her the items will reach Haradheere by nightfall the following day. After agreeing upon a price, Hibaaq writes down an account number and tells the security guard at the gate she is heading to Dahabshiil to make a money transfer and he should expect her back in an hour. If she does not return within that time, he should immediately notify her sister who will dispatch instructions to the private security company they have on payroll as well as the head pirates overseeing each camp.

Hibaaq, the financial investor, is an example of a woman in a contemporary piracy leadership position. Popular culture's wild imaginaries of women pirate leaders in paperbacks, films, and television series discussed in the previous chapter make no mention of women who manage piracy's quotidian business operations. The women pirate leaders of popular culture are constantly swept up in dramatic battles and love affairs, yet the realities of contemporary women pirate leaders' lives are much closer to the description offered here of a Somali woman managing operational finances from a backroom desk. These popular cultural heroines command fleets of ships, enlist captives, and gracefully employ their swords in battle. Although seemingly pure fantasy, these popularized images loosely derive from a small yet historically significant group of real women pirate leaders who led and fought alongside their male counterparts on the high seas.

Historical accounts of women pirate leaders date to as early as the third century BC. These extraordinary women shared a similar set of pathways into their roles of pirate leadership. Most took over leadership positions upon the deaths of their pirate husbands or fathers, while other women became pirate leaders through marriage. The advantages that nearly all these women shared were that they were associated with—through marriage or bloodlines—men who had already gained the respect of all-male pirate crews.

Plundering by Marriage

One of the earliest known female pirate leaders is Queen Teuta of Illyria (present-day Albania). Piracy was commonly practiced along Illyria's coast, especially in the Adriatic Sea, when Teuta rose to power in 231 BC after the death of her husband, the king. Upon assuming leadership of the kingdom, Queen Teuta encouraged her subjects to plunder and raid Greek and Roman ships to enrich Illyria and maintain their country's independence from Greek, and later Roman, colonization (Faniko & Karamuco, 2016). Historians remain unsure whether Teuta "led from a distance" or commanded any of the pirate fleets at sea. Nevertheless, her leadership and support of piracy was instrumental in developing and securing Illyrian independence.

Like Queen Teuta, Aethelflaed (c. 870–918) assumed power upon her husband's death and became the first woman to rule an Anglo-Saxon kingdom. She became a fierce military leader during a period when Danish Vikings were pillaging and otherwise controlling most of England. Described in historical documents as "kingly" and a "war leader", it is more probable that Aethelflaed commanded the fleet from shore although her husband had actively commanded pirate vessels at sea prior to his death (Stafford, 2017: 103). Her leadership, including her involvement in piracy, played a central role in reshaping the political geographies of Northern Europe when her forces defeated Viking raiders and Danish pirates to reclaim English lands.

Queens Teuta and Aethelflaed's pathways into piracy leadership involved assuming political positions formerly held by their deceased husbands. However, politics could influence women pirates' leadership in other ways as well, as occurred in the case of well-educated, wealthy French noblewoman turned pirate Jeanne de Clisson (1300–1359). There is no evidence that her husband had been involved in piracy, and de Clisson only became a pirate to seek vengeance for her second husband's execution for treason by the French king. Following his execution, de Clisson denounced her country, raised an army of her husband's loyalists, and set sail to aid in the attack of French ships and the slaughter of French citizens. Her legendary command of three warships and her brutal treatment of captive sailors earned her the nickname, "The Lioness of Brittany" (O'Brien, 2022).

de Clisson's strategic use of piracy as a political tool is characteristic of the time period, as prior to the 1800s territorial expansion and resource conflicts often played out at sea. Governments would contract out private ships, which the British termed privateers, to destroy rival commercial shipping vessels to establish mercantile superiority and naval power (Hillmann & Gathmann, 2011). Sayyida al-Hurra (1485–1561), a celebrated Muslim heroine and successful pirate queen, a title that recognizes the intersection of illicit activities and power, became governor of Tetouan, a coastal state in Morocco following her husband's death. al-Hurra allied herself with another famous Turkish male pirate and assembled a fleet of ships, capturing Spanish and Portuguese ships and ransoming Christian prisoners across the Mediterranean. Her strategic use of piracy helped solidify Tetouan as a regional power and prosperous city built on the spoils of piracy raids (Karim et al., 2021).

Women pirate leaders such as Queens Teuta and Aethelflaed, Jean de Clisson, and Sayyida al-Hurra are much lesser known than Anne Bonny and Mary Read, who are arguably the most identifiable female pirate leaders. Anne Bonny (1697–1721) was born to unmarried parents, a considerable stigma at that time in Ireland and later taken by her father to America where they settled in South Carolina. Historical accounts of Bonny's life describe her as rebellious, fierce, and promiscuous—the opposite of gender exceptions of the time that encouraged women to be passive, submissive, and monogamous (Defoe, 1724/2012). She purportedly ran away with a man to the Caribbean where they both joined a band of pirates that included the famous Calico Jack Rackham, who became Bonny's lover.

Bonny's path intersected with Mary Read's when Rackham's crew conquered a ship and held Read who was among the hostages. Read is also described as a woman who wore men's clothing to gain employment as a sailor and escape the poverty and social restrictions of life on land. Accounts vary, but some claim that Bonny and Read were the first openly lesbian pirates (Garber, 2021). Pirates, particularly those in the buccaneer community of Bonny's and Read's era, did not adhere to then-prevalent taboos against same-sex sexual relationships (Burg, 1995; Murray, 1992), so a sexual relationship between Bonny and Read would not be unfathomable or abnormal given the context. What is notable and problematic, however, is the entanglement of the women's sexual preferences and activities with details of their reputations as tough, ruthless pirates. It is unsurprising, then, that these two women often serve as the inspiration for the hypersexualized female pirate imagery promoted today for entertainment purposes.

In sharp contrast to the dramatic, sexualized imaginaries of Bonny and Read and much more akin to Hibaaq's administrative role in Somalia, one of the most successful pirates in history is Zheng Yi Sao, also known as Cheng I Sao or Ching Shih (1775–1844). Born in a Chinese brothel inn, Soa was a sex worker who married the commander of a fleet of pirate vessels that would become the largest pirate confederation in history (Murray, 1981). She honed her sailing, fighting, and leadership skills with her husband and upon his death assumed leadership of his fleet estimated to include 2,000 ships and over 70,000 men, and defeated maritime superpowers including China, Portugal, and England, solidifying her domination of the South China Sea (Murray, 2004). Sao is known for imposing one of the strictest piracy codes, which included punishments such as immediate decapitation for disobeying superiors and committing capital offenses such as murdering an unarmed civilian (Murray, 1987).

All these fierce women leaders found their way into piracy through marriage or relationships with male lovers. Whereas some joined and pirated alongside their husbands from the beginning, others were bequeathed the role of pirate leaders upon their husband's death or claimed it in revenge. These intimate relationships provided a pathway into piracy leadership roles at a time when law and society regarded women as men's property. Some of these women assured their crew's obedience out of respect for their late husbands or lovers, while, for others, obedience was earned, challenged, and held on to daily through their actions.

Next-of-Kin and Parvenu Pirates

The father–daughter relationship also proved to be a promising pathway into piracy leadership, particularly in Northern and Western Europe. During the Viking Age (793–1066 AD), Scandinavian women regularly participated in the raiding, colonization, and conquest of neighboring areas of Europe alongside men. One of the most famous of these women is Ladgerda (ninth century AD), a Danish Viking pirate known as a shieldmaiden. Both historical and mythological accounts of Ladgerda describe her as a skilled female fighter with a fierce temper who battled at the front lines (Clover, 1986). As a king's daughter and sole surviving heir, she was born into leadership and ruled over what is today Norway. Her noble bloodline and battle-hardened reputation caught the eye of Danish pirate king, Ragnar Lodbrok. The two married and spent many years pirating and conquering villages along the North Sea. Although she married a pirate, her battle experience and leadership role are credited to her father. Like many historical accounts, women pirates are depicted as "receiving" their skills from husbands, fathers, or male intimate partners, rather than independently cultivating these skills.

Lady Mary Killigrew and Grace O'Malley, like Ladgerda, came from privileged lineages and circumstances. Both women's families had histories of pirating during Europe's early modern period, early to mid-1500s, when King Henry VIII was pushing to expand England and conquer surrounding territories. Pirating was a means to fend off these royal conquests in the absence of a strong military while providing pirate leaders with the opportunity to enrich themselves among ever-shifting geopolitics. Both Killigrew and O'Malley pirated along the English coast with their fathers during this era and continued after their fathers' deaths. The women successfully commanded fleets and battled rival clans of pirates to keep their coastal communities safe during a period of great maritime unrest and violent land grabs (Burwick & Powell, 2015).

Killigrew and O'Malley's fathers viewed them as capable fighters and promising leaders and raised them with expectations of learning and mastering the intricacies of combat and pirating necessary to defend their families and homelands. It is no surprise that these women eventually wed fellow pirates, as they lived their lives immersed in battles and surrounded by pirates since birth. However, not all women pirate leaders had husbands, lovers, or fathers involved in piracy. A small number of women pirate leaders appear to be self-made, entering piracy of their own free will, including the aforementioned Mary Read and Liu Laijiao of China whose name is associated with other menacing and formidable pirates operating in the South China Sea during the Qing dynasty.

Facilitators, Careworkers, and Entrepreneurs: Underestimating "Assistant Roles"

Barkhado crouches down and sets her torch on the ground in the center of the dark room. The sun is slowly rising but the light is insufficient to aid her aging eyes. She opens the notebook and begins counting the guns and boxes of ammunition stacked against the wall. As she moves slowly and meticulously from left to right, her phone rings. She recognizes the number as that of her cousin who recently passed away. Two years earlier, when she was living in Garowe, married to a war-traumatized man more committed to chewing khat than caring for his wives and children, he approached her with a business opportunity. Barkhado had already established a small business selling sweets and baked goods from a table on the roadside to keep the family fed. Word of her husband's struggles and her keen business intuition quietly and quickly spread through the family networks and her cousin paid a visit.

"I have a very lucrative opportunity for you", he said, "but it will require you to relocate to Bosaso and cut ties with your husband. I will see to it that your children are well fed, their school fees paid, and they have opportunities for a better life". Barkhado did not need much convincing, and the next morning she left with her cousin for Bosaso. The cinderblock compound she has been occupying since that day sits just on the edge of town. Not too far from the Port of Bosaso, it provides easy access to the city but is far enough removed that only those who need to find her take the time to venture out beyond the mass of refugee aqals (makeshift huts). She lives in the front room and stores the guns and ammunition she sells to pirates in the back room. Covered trucks bring in new supplies and transport out recently purchased ones under the cover of darkness. She is aware of the rumors and judgment due to her involvement with pirates, but she is proud of her decision to make a better life for herself and her children. When she answers the phone, her late cousin's wife, Hibaaq, is on the other end wanting to place an order.

Perhaps the largest contribution women make to piracy operations is the labor they perform as facilitators, careworkers, and entrepreneurs. Despite most of these roles being shore based, they are vital to the success and longevity of piracy operations. Women have long assisted pirates by providing safe refuge in their homes and inns, meals to nourish their bodies after a long journey at sea, escape and safe passage when law enforcement comes knocking on doors, and, more recently, selling guns and providing seed money for piracy operations. Some of these roles capitalize on existing gendered cultural expectations of women as natural domestic workers. As a result, specialized women-led cottage industries have emerged and evolved to cater specifically to the unique needs of piracy operations while also taking advantage of the law's blind spot for women as perpetrators of crime.

The under-theorization of women as crime bosses and leaders also extends to their participation in other roles important to the operation and expansion of criminal enterprises (Belknap, 2020). The criminal justice system's treatment of women's participation in organized crime tends to spare them harsher penalties in ways that reflect cultural beliefs that women lack the competencies and intellect to play a real role in organized crime. For example, judges regarded women Italian mafia "assistants" as passive subjects who were either not smart enough to launder money or their roles were not sophisticated enough to be punishable by law (Longrigg, 1998). These stereotypes are widespread in criminal justice practitioners' underestimation of women's ability to comprehend complex business transactions and other entrepreneurial tasks related with illicit and illegal activities. When women are caught, their activities are somehow deemed "not criminal enough" (or violent enough) to warrant criminal–legal sanctions (Arsovska & Allum, 2014). The architecture of gendered ignorance surrounding women's participation in organized crime is instrumental in enabling them to use, capitalize on, and enhance their unique skillsets (Dewey et al., 2019). This extends to piracy, where facilitators and entrepreneurs continue to operate in a climate where organized crime leaders and criminal justice professionals alike regard women's participation as fringe and less important than their male counterparts.

Pirates operating at sea have long been romanticized, fetishized, and prioritized in terms of policy responses and law due to the threats they pose to capital and public safety. The action of stalking ships, the exchange of gunfire between vessels, boarding the vessel, and taking the crew hostage—all these activities take place on the water. Most of these crimes take place out of sight of the public and even law enforcement, yet piracy scholars and antipiracy practitioners alike prioritize what happens *at sea* rather than the considerable onshore efforts to obtain financing and other necessary aspects of a piracy operation. Although the United Nations Convention on the Law of the Sea (UNCLOS) expanded its definition of piracy in response to the early twenty-first-century Somali piracy crisis to include "any act of inciting or of intentionally facilitating an act described" to allow for the arrest and prosecution of those on shore that aid in piracy operations, some fear that expansion may interfere with state sovereignty (United Nations, 2022; Gaynor, 2012). This presents a conundrum. Either accept that piracy is a shore-based crime that manifests at sea and adopt state laws reflecting this, or only identify and prosecute piracy activities that occur at sea and use existing state laws—such as those against money laundering, murder, and kidnapping—to prosecute the shore-based activities (Chalk & Hansen, 2012).

The epistemic divide between piracy understood as a sea-based crime versus a land-based crime that manifests at sea directly informs who police identify as pirates and the related policies and programs aimed at pirates' arrest and prosecution. This divide may account for the long-standing treatment and analysis of shore-based activities as *peripheral* to piracy operations rather than *central* to them. A successful pirate attack may entail a mother ship and a few smaller attack skiffs making frequent trips out to sea or staying at sea for days before the apprehension of a vessel. However, tasks such as finding an investor, obtaining weapons, building a base camp, servicing and maintaining the base camp, guarding and caring for hostages, negotiating a ransom, washing ransom money, moving goods, and planning subsequent attacks involve an extensive network of people on shore. The attack at sea is just one brief aspect of a much larger geographically expansive land-based piracy operation carried out over a period of months to years. When this network is made visible, so too are the many important contributions made by women.

The remaining sections weave together examples of women's central roles in shore-based piracy operations. These examples demonstrate overlap and continuity between historical periods and across geographical locations while also presenting some contextually unique roles. What emerges is evidence of a network of entrepreneurial women, whose service is vital to the maintenance and success of maritime piracy.

Facilitators

Piracy is a highly mobile form of organized crime requiring continuous movement of people and resources at sea, along the coastline, and even inland. Pirates operating during the 1700s in the Caribbean had well-established points of contact at various ports along the shores of North American colonies, England, Ireland, Africa, and the South China Sea (Hurd, 1925; Rediker, 2004; Defoe, 2012). As they moved from port to port, their contacts, many of whom were women, provided them with shelter and nourishment and moved their plundered goods while in port. These "pirate nests", as pirate-friendly ports were known, received great wealth in return for their support to pirates (Hanna, 2015). One governor-general dispatched to eradicate piracy along the Guangdong coast of South China even relayed to Emperor Qing that the only reason piracy was flourishing was because people on the coast harbored pirates by giving them aid and selling their plundered goods (Antony, 1992). Support on land has been and continues to be a critical lifeline for successful piracy operations.

Inns, alehouses, brothels, and hotels frequented by pirates provide more than rest and respite. In seventeenth-century England, ale houses and lodging houses not only provided refuge for pirates while onshore, but they were also critical to the disposal, movement, and selling of pirated cargo (Appleby, 2013). These facilities were often operated by sailors' wives or single women familiar with seafarers' precarious and mobile lifestyle. In the context of Chinese piracy, many of the women helping pirates to sell stolen goods or launder money had ties to piracy through family and used their networks to their financial advantage (Kwan, 2020). Women's ability to move between the criminal and noncriminal worlds of crowded, bustling ports with an ear to the ground also positioned them as pirates' key informants. These "middle(wo)men" were vital to the movement of material items and exchange of important information that helped pirates evade arrest and identify upcoming shipping movements and potential prey (Kwan, 2020: 205).

Women's participation as aiders/informants (facilitators) in maritime piracy is similar to some of the roles women play in drug trafficking. Anderson's (2005) research on the illicit drug economy unveils how women's ability to secure and maintain a home as a base to provide basic sustenance and social support is fundamental to the drug world's organization and accumulation of capital. Women's household maintenance provides male dealers who occupy higher positions in the drug economy with the time to obtain or manufacture and (re)sell illicit drugs, accumulate capital more quickly, and attain structural power in the market. In turn, the female heads of household gain social and financial capital for themselves by providing the shelter and nourishment needed for the dealer to thrive (Sommers et al., 2000).

The overlap between the middle-women of piracy and drug dealing underscores the importance of physical stability, time, and situated knowledge for capital accumulation. Both criminal enterprises require safe harbor and someone who knows the intimate ins and outs of the local market in order to move goods and achieve financial gain. Women have stepped in to meet these needs and fulfill transactional roles in both piracy and drug trafficking. While exercising agency, they are also able to gain increased social and financial capital from the arrangements and rely upon their gendered criminal invisibility to eschew law enforcement. State approaches to both piracy and the drug trade have backfired in similar ways. For example, backdoor deals to pay ransom for piracy hostages save face for states but still enrich pirates. Similarly, drug programs aimed at alternatives to incarceration too often seek to channel women's agency and sense of empowerment into benefiting others (e.g., becoming effective mothers) rather than finding ways for them to use their empowerment and agency for their own fulfillment outside of the world of drugs (Anderson, 2005).

Entrepreneurs

Piracy produces entrepreneurial opportunities for those who can carve out a niche for themselves. Three fascinating examples of business ventures established by women to serve the operational needs of piracy in historical and cultural contexts include recruiting and protection rackets, resource dealers, and financial investors. Recruiting and protection rackets were more common historically, particularly in Chinese piracy. Women as resource dealers and financial investors are most evident in case studies from contemporary piracy off the coast of Somalia.

Women involved in Chinese piracy were instrumental in recruiting pirates and supervising protection rackets, two activities that ensured a continual surplus of labor at sea along with a steady stream of income to support operations during periods of low piracy activity. Women also approached members of the local militia and attempt to recruit them into piracy by offering them rice and even money as a signing bonus (Kwan, 2020). Much of this recruitment took place across the various inland areas of China where local militia members with combat or military experience were desperate for work. The women worked diligently onshore to ensure there was never a shortage of young, well-trained male bodies ready to fight.

Chinese women also supervised protection rackets that promised paying seafarers and coastal businesses safety from pirate attack, a form of extortion common in organized crime. These protection rackets brought in a steady amount of money that, at times, was more than the amount of money pirates could acquire through shipboard raids (Kwan, 2020). They set up financial offices along the coast where clients could pay their protection fees, produced and distributed protection papers to paying clients, extracted taxes from shipping vessels, oversaw the sale of captives into slavery, and managed the collected treasury from these ventures. Notable women who oversaw these types of rackets, such as Mrs. Bigfoot, Ng Akew, and Liu Laijiao, became rich alongside enriching their piracy enterprises (Kwan, 2020).

Contemporary piracy also provides examples of women running impressive financial operations that provide the lifeline for piracy enterprises. Ethnographic studies conducted in Somalia reveal that women assist with the daily maintenance of piracy operations by collecting and distributing goods such as food, clothes, khat, petrol, and guns to pirates. Former hostages recall being kept alive on food that was prepared and delivered by local women who own restaurant businesses (Chandler et al., 2012; Moore, 2018) and members of the Somali public report knowing women who lease guns and sell petrol to pirates along the Bosaso coastline (Gilmer, 2019).

Somali gendered cultural expectations that appear restrictive from the outside, such as limited courting under the watchful eye of relatives, created a breadth of opportunities for Somali women to demonstrate agency and benefit financially from piracy. As piracy off the coast of Somalia peaked around 2009–2012, piracy groups became increasingly reliant on women to carry out the feminized activities of purchasing and transporting goods needed to sustain a piracy operation (Gilmer, 2014, 2019). Stories of women like Barkhado leaving their villages and families in Somalia's interior to set up businesses along the coast near piracy bases became increasingly common. Through these ventures, women were purportedly making more money than they did in the years following the onset of the Somali Civil War (1991–present) when they became more active in business and engaged in small-scale trade, selling clothes, vegetables, snacks, and household items in the market (Ingiriis & Hoehne, 2013).

Like the Chinese women running protection rackets, some Somali women demonstrated keen financial prowess in their roles as piracy investors.

Details of these women's lives remain relatively unknown, but the impact of their investments cannot be understated. Throughout history, establishing and maintaining a piracy operation has been an incredibly costly venture. Seed money is required to purchase boats, guns, and nourishments before the operation commences. As the operation gets underway, additional funding is needed for the continual replenishment of supplies. In the context of contemporary piracy off the coast of Somalia, holding hostages for months to years requires a substantial amount of initial and ongoing financial investment until a ransom is paid. Therefore, piracy investors play a critical role in the existence and continued operation of modern piracy. With Somalia remaining one of the poorest countries in the world, a relatively small number of individuals have the capacity to serve in the role of investors—most of which have included Somalia's elite or members of the Somali diaspora (Dua & Menkhaus, 2012; Percy & Shortland, 2013; World Bank, 2022).

The existence of women piracy investors has been verified by American and British military personnel and diplomats based in Kenya and working on counter-piracy operations. Intelligence officials shared with me that one of the investors of the largest piracy operation based near Haradheere was, indeed, a woman who took over the business when her husband died. Others noted they were aware of two women investors operating out of the Galmudug region of Somalia. Although these women exist, their role as piracy investors is considered an exception to the norm. Research suggests that the main financiers of piracy in Somalia are elder (male) pirates who had participated in piracy during the early 1990s (Hansen, 2009; Bahadur, 2011). Which begs the question of how a handful of Somali women came to occupy this role. Those familiar with the women claim they either inherited the money from the death of a husband and then decided to become involved in piracy, or their husbands were already involved in piracy investing and the women took over for them after their deaths—reinforcing relationship and kinship pathways as a continuity throughout time and across regions.

Exploitation and Edgework

Research on women's roles in transnational organized crime provides a starting point to theorize a general framework for understanding women's paths to leadership roles in piracy. Whereas piracy has liberatory potential for traditionally marginalized groups such as poor and working-class people, including black men (Kinkor, 2001; Bromley, 2001) and queer-identified people (Turley, 1999), contemporary studies on women's roles in organized crime, human trafficking, and drug trafficking groups provide fertile ground for rethinking women's pathways into piracy roles beyond the traditional tropes of emancipation and liberation (Simon, 1975; Adler, 1975).

Groundbreaking qualitative research on the Neapolitan Camorra, an Italian mafia-type criminal organization, can help us rethink how operational contexts and structure can create opportunities for women in organized crime. According to Allum and Marchi (2018), the Camorra predominantly resides and operates in the most chaotic and deprived districts of Naples where crime and poverty are prevalent. This environment, along with the Camorra's fragmented structure and loose conception of family, converges and creates spaces of entry for women in the highest levels of the organization. Perhaps, a similarly chaotic and lawless space such as the high seas, coupled with the low barrier of entry for piracy beyond bravery and a willingness to contribute to the mission, fight when called upon, and abide by the codes of the ship, opened a whole new working environment for women (Urbina, 2019). Similarly, contemporary piracy in failed states such as Somalia relies on an extensive network of lineages and clans for entry. At the most basic level, maritime piracy shares some fundamental characteristics with the Camorra in that a chaotic and impoverished environment and flexible membership structure enabled women to enter the higher echelons of the organizations.

Camorra women tend to enter leadership roles when situational factors require the clan to exploit available resources and skills—women. Allum and Marchi (2018) refer to this Marxist labor-based phenomenon as a "call to arms" of the Camorra's "female reserve army" (p. 369). The wives, mothers of children, and daughters of the Camorra leaders having intimate knowledge of the inner workings of the organization assume leadership role in "times of crisis" (e.g., imprisonment, hiding, or death) when a male successor is not available. The women earn their status and social capital via their relationship, whether it be romantic or bloodlines, to the leader. The members of the organization respect and obey these women as proxies of the male leader in his absence. This "call to arms" phenomenon is also seen in women pirates' rise to leadership in times of crisis. Historically, the "time of crisis" for pirate groups was spurred by the death of a captain or other leading figure. Whether it was Zheng Yi Sao taking control of the largest pirate fleet in history or Ladgerda ravaging the shores of Northern and Western Europe, these women assumed control immediately following the deaths of their lovers, husbands, or fathers. Their close association with the deceased helped garner a level of respect and obedience among the crew that enabled the group to carry on and move forward with their expeditions under the direction of women leaders. As early apprentices to their loved ones, the women pirates were not only a "reserve army" of sorts, but some had been serving on active duty alongside their male counterparts prior to occupying leadership roles (Allum & Marchi, 2018).

The small number of women who have served as leaders and fighters in male-dominated maritime piracy underscores their rarity and further supports an understanding of women occupying these roles as a means to obtain and exploit resources rather than an emancipatory norm. However, what about the self-made women leaders and fighters who did not come into piracy because of prior associations with male pirates? Perhaps, we can better understand women's involvement in piracy as a form of maritime edgework. The concept of edgework was developed by Lyng (1990) to theorize why some individuals engage in risk-taking behavior such as skydiving and bungee jumping. These high-risk pursuits require both thrill and skill as participants negotiate the boundary between order and chaos, safety and harm, and socially acceptable and unacceptable practices (Lyng, 2005). Critical feminist scholars have pushed the edgework model to acknowledge the gendered experiences of edgework. Historically, women who engage in high-risk activities were viewed as mentally ill. However, women do engage in edgework like men. They push their bodies, they experience a rush, and they use their skills to negotiate the edge and avoid harm or death. Examining women doing edgework expands our understanding of "risks" and "skills" to include those beyond the physical body.

Edgework has also been reconceptualized as a form of resistance to violence and patriarchal control. Women as an oppressed class are far more likely to exist in a space of high risk and potential violation than men (Stanko, 1997). The misogyny underlying that high risk equates performing femininity with keeping oneself safe. Thus, the edge becomes a space where women can seek out fearful experiences to empower themselves; they can confront normative expectations that women are risk adverse and fearful (Olstead, 2011). Women can find rewards in the form of power and control over their oppressors, and in some case their abusers, through their skillful and successful navigation of harms at the edge (Rajah, 2007). But what is the edge?

The shifting boundaries of the edge are co-constitutive of social and cultural symbolizers and signifiers of risk. The edge can be an activity, a space, a place—anything that signifies risk. Thus, the edge and edgework are continually being reformulated. What constituted edgework in the past may look very different from what constitutes edgework in contemporary times and across contexts. Maritime spaces have long been viewed as high-risk spaces, particularly for women. Women piracy leaders and fighters were not only operating at the physical edge of society—the high seas—but they were also occupying and performing roles considered high risk and unacceptable for the period—commanding men and engaging in "dirty work" such as violence and murder. The historical accounts of these women lusting after various men and embracing their sexuality also suggest that they were perceived as existing on the affective edges defined in opposition to the chaste and heteronormative expectations of society.

These women chose to engage in piracy activities despite the observable threats to their physical or mental well-being and place in an ordered society that aligns with the very definition of edgework (Lyng 1990). The edgework fulfilled an unmet need, perhaps a sense of control over one's life or the desire for challenge and excitement. Through their apprenticeships and true grit, the women armed themselves with the skills needed to live and work at the edge. Pirates were already viewed as social outcasts operating on the wild untamed waters of the sea. Women leaders and fighters of historical piracy further reconstituted that edge in ways that society is yet to see replicated in contemporary times. Arguably, the shift from piracy as an "escape society" and evasion of state oppression to the present piracy as an economic venture model may have altered the way women think about the risk and thrills of piracy activities at sea (MacKay, 2013). As women's roles in piracy continue to evolve, so too does the constitution of edgework as the *edge* of piracy moves from offshore onto land.

The examples discussed in this chapter both challenge and support the "gendered markets" perspective of organized crime that identifies a gender-based division of labor within criminal groups (Zhang et al., 2007). Women have long capitalized on the various labor niches (Campbell et al., 2001; Savona & Natoli, 2007) of piracy to reap the financial benefits from shore while avoiding the violence and high-risk physical demands of leadership and fighter roles at sea. Some of these niches may be explained through a gendered division of labor lens. For example, the women of early English piracy who aided pirates by providing them with shelter, food, and sexual gratification in turn for money. Similarly, the resource dealers of Somali piracy capitalized on their government's insistence that women are not viewed as pirates or pirate accomplices to safely distribute food, water, petrol, and other essentials to the pirate groups. However, the women running the recruitment and protection rackets of Chinese piracy and the Somali women investing seed money in piracy operations are not convincingly occupying these roles due to their gender, but rather due to their intellect and advanced skills.

While gender may present unique opportunities for women's involvement in piracy and other organized crime, it is not the only explanation. Miller (2002) insists on acknowledging the complexities of agency and social practice in a way that strikes a balance between recognizing the significance of gender but not reducing everything to gender. Accordingly, during the Golden Age of Caribbean and Western European piracy, women often broke with traditional gender roles and took over the business responsibilities of men who had set sail for extended periods of time (Appleby,

2013). The actions of women who bought pirated goods, fed pirates, and even acted as pawnbrokers for pirates may be better understood as a general blurring of boundaries relating to gendered expectations as well as legal/illegal activities in the context of rapid change and economic upheaval resulting from conflict, war, and territorial expansion.

The next chapter reveals the more intimate economy of piracy by exposing the roles of women working in piracy-adjacent trafficking markets. Heeding the call from Siegel and Turlubekova (2019), Chapter 3 analyzes women's roles in the context of cross-border transnational activities, new markets, and international migrations. Accordingly, it engages with women's increasing involvement in the contemporary ransom piracy model, how their roles intersect with new and old patterns of globalized labor exploitation and migration, and how these roles manifest in the structure of piracy in Somalia.

References

Adler, F. (1975). *Sisters in crime: The rise of the new female criminal*. New York: McGraw-Hill.

Allum, F., & Marchi, I. (2018). Analyzing the role of women in Italian mafias: The case of the Neapolitan Camorra. *Qualitative Sociology*, *41*(3), 361–380.

Anderson, T. L. (2005). Dimensions of women's power in the illicit drug economy. *Theoretical Criminology*, *9*(4), 371–400.

Antony, R. J. (1992). The suppression of pirates in South China in the mid-Qing period. *American Journal of Chinese Studies*, *1*(1), 95–121.

Appleby, J. C. (2013). *Women and English piracy, 1540–1720: Partners and victims of crime*. Rochester: Boydell & Brewer Ltd.

Arsovska, J., & Allum, F. (2014). Introduction: Women and transnational organized crime. *Trends in Organized Crime*, *17*(1–2), 1–15.

Bahadur, J. (2011). *The pirates of Somalia: Inside their hidden world*. Visalia: Vintage.

Belknap, J. (2020). *The invisible woman: Gender, crime, and justice*. Thousand Oaks: SAGE Publications.

Bromley, J. S. (2001). Outlaws at sea, 1660–1720: Liberty, equality and fraternity among the Caribbean freebooters. In C. Pennell (Ed.), *History from below: Studies in popular protest and popular ideology* (pp. 293–318). New York: NYU Press.

Burg, B. R. (1995). *Sodomy and the pirate tradition: English sea rovers in the seventeenth century Caribbean*. New York: NYU Press.

Burwick, F., & Powell, M. N. (2015). *She-Pirates: British pirates in print and performance*. New York: Palgrave Macmillan.

Campbell, A., Muncer, S., & Bibel, D. (2001). Women and crime an evolutionary approach. *Aggression and Violent Behavior*, *6*(5), 481–497.

Chalk, P., & Hansen, S. J. (2012). Present day piracy: Scope, dimensions, dangers, and causes. *Studies in Conflict and Terrorism*, *35*(7–8), 497–506.

Chandler, P., Chandler, R., & Edworthy, S. (2012). *Hostage: A year at gun point with Somali pirates*. Chicago: Chicago Review Press.

Clover, C. J. (1986). Maiden warriors and other sons. *Journal of English and Germanic Philology, 85*(1), 35–49.

Defoe, D. (1724/2012). *A general history of the pyrates*. Courier Corporation.

Dewey, S., Zare, B., Connolly, C., Epler, R., & Bratton, R. (2019). *Outlaw women: Prison, rural violence, and poverty on the new American west*. New York: NYU Press.

Dua, J., & Menkhaus, K. (2012). The context of contemporary piracy: The case of Somalia. *Journal of International Criminal Justice*, *10*(4), 749–766.

Faniko, I., & Karamuço, E. (2016). The Albanian country, the institutional and territory organisation in the Illyrian period and during the principality of "Arbëria". *Mediterranean Journal of Social Sciences*, *7*(Suppl. 1), 21.

Fiandaca, G. (Ed.). (2007). *Women and the Mafia: Female roles in organized crime structures* (Vol. 5). Berlin: Springer Science & Business Media.

Garber, L. (2021). Unsafe seas for women. In L. Garber (Ed.), *Novel approaches to lesbian history* (pp. 111–144). Cham: Palgrave Macmillan.

Gaynor, J. (2012). Piracy in the offing: The law of lands and the limits of sovereignty at sea. *Anthropological Quarterly*, *85*(3), 817–858.

Gilmer, B. (2014). *Political geographies of piracy: Constructing threats and containing bodies in Somalia*. Berlin: Springer.

Gilmer, B. (2019). Invisible pirates: Women and the gendered roles of Somali piracy. *Feminist Criminology*, *14*(3), 371–388.

Hanna, M. G. (2015). *Pirate nests and the rise of the British empire, 1570–1740*. Chapel Hill: UNC Press Books.

Hansen, S. J. (2009). *Piracy in the greater Gulf of Aden*. Oslo: Norsk institutt for by-og regionforskning.

Hillmann, H., & Gathmann, C. (2011). Overseas trade and the decline of privateering. *Journal of Economic History*, *71*(3), 730–761. Retrieved from http://www.jstor.org/stable/23018337

Hurd, A. (1925). *The reign of the pirates*. New York: Alfred A. Knopf.

Ingiriis, M. H., & Hoehne, M. V. (2013). The impact of civil war and state collapse on the roles of Somali women: A blessing in disguise. *Journal of Eastern African Studies*, *7*(2), 314–333.

Karim, A. A., Al-Jundi, H., & Khalil, R. (2021). Female pioneers in Islamic Middle Ages: A theological and psychological perspective. In A. Karim, R. Khalil, A. Moustafa (Eds.), *Female pioneers from ancient Egypt and the Middle East* (pp. 29–45). Singapore: Springer.

Kinkor, K. J. (2001). Black men under the black flag. In R. Pennell (Ed.) *Bandits at sea: A pirates reader* (pp. 195–210). New York: NYU Press.

Kleemans, E. R., Kruisbergen, E. W., & Kouwenberg, R. F. (2014). Women, brokerage and transnational organized crime: Empirical results from the Dutch organized crime monitor. *Trends in Organized Crime*, *17*(1), 16–30.

Kwan, C. N. (2020). In the business of piracy: Entrepreneurial women among Chinese pirates in the mid-nineteenth century. In J. Aston & C. Bishop (Eds.), *Female entrepreneurs in the long nineteenth century* (pp. 195–218). Cham: Palgrave Macmillan.

Longrigg, C. (1998). *Mafia women*. Bethesda: Arrow.

Lyng, S. (1990). Edgework: A social psychological analysis of voluntary risk taking. *American Journal of Sociology*, *95*(4), 851–886.

Lyng, S. (2005). Edgework and the risk-taking experience. In *Edgework: The Sociology of Risk-Taking* (pp. 3–16). New York: Routledge.

MacKay, J. (2013). Pirate nations: Maritime pirates as escape societies in late Imperial China. *Social Science History*, *37*(4), 551–573.

Miller, J. (2002). The strengths and limits of 'doing gender' for understanding street crime. *Theoretical Criminology*, *6*(4), 433–460.

Moore, M. S. (2018). *The desert and the sea: 977 days captive on the Somali pirate coast*. New York: Harper Wave.

Murray, D. (1981). One Woman's Rise to Power: Cheng I's Wife and the Pirates. *Historical Reflections/Réflexions Historiques*, 147–161.

Murray, D. H. (1987). *Pirates of the South China coast, 1790–1810*. Stanford, CA: Stanford University Press.

Murray, D. H. (1992). The practice of homosexuality among the Pirates of late 18th and early 19th century China. *International Journal of Maritime History*, *4*(1), 121–130.

Murray, D. H. (2004). Piracy and China's maritime transition, 1750–1850. In G. Wang & C.-K. Ng (Eds.), *Maritime China in transition 1750–1850* (pp. 43–60). Wiesbaden: Harrassowitz.

O'Brien, E. (2022). Historical fiction and the Breton landscape: Writing the life of Jeanne de Belleville. *Life Writing*, 1–15.

Olstead, R. (2011). Gender, space and fear: A study of women's edgework. *Emotion, Space and Society*, *4*(2), 86–94.

Percy, S., & Shortland, A. (2013). The business of piracy in Somalia. *Journal of Strategic Studies*, *36*(4), 541–578.

Rajah, V. (2007). Resistance as edgework in violent intimate relationships of drug-involved women. *British Journal of Criminology*, *47*(2), 196–213.

Rediker, M. (2004). *Villains of all nations: Atlantic pirates in the golden age*. Brooklyn: Verso.

Savona, E., & Natoli, G. (2007). Women and other mafia-type criminal organizations. *Women and the Mafia*, 103–106. New York: Springer.

Siegel, D., & De Blank, S. (2010). Women who traffic women: The role of women in human trafficking networks–Dutch cases. *Global Crime*, *11*(4), 436–447.

Siegel, D., & Turlubekova, Z. (2019). Organized crime in Kazakhstan. In *Organized crime and corruption across borders* (pp. 166–182). Oxfordshire: Routledge.

Simon, R. J. (1975). *Women and crime*. Lexington, MA: Lexington Books.

Sommers, I. B., Baskin, D. R., & Fagan, J. (2000). *Workin'hard for the money: The social and economic lives of women drug sellers*. Hauppauge: Nova Publishers.

Stafford, P. (2017). The annals of Æthelflæd': Annals, history and politics in early tenth-century England. In J. Barrow & A. Wareham (Eds.), *Myth, rulership, church and charters* (pp. 115–130). Oxfordshire: Routledge.

Stanko, E. A. (1997). Safety talk: Conceptualizing women's risk assessment as a technology of the soul. *Theoretical Criminology*, *1*(4), 479–499.

Tregunna, A. (2014). Cocaine cowgirl: The outrageous life and mysterious death of Griselda Blanco, the godmother of Medellin. *Trends in Organized Crime*, *17*(1), 132–134.

Turley, H. (1999). *Rum, sodomy, and the lash: Piracy, sexuality, and masculine identity*. New York: NYU Press.

United Nations. (2022). United nations convention on the law of the sea. Retrieved from https://www.un.org/depts/los/convention_agreements/texts/unclos/unclos_e.pdf.

Urbina, I. (2019). *The outlaw ocean: Journeys across the last untamed frontier*. Visalia: Vintage.

World Bank. (2022). Retrieved from https://data.worldbank.org/indicator/NY.GDP.PCAP.CD?most_recent_value_desc=false.

Zhang, S. X., Chin, K. L., & Miller, J. (2007). Women's participation in Chinese transnational human smuggling: A gendered market perspective. *Criminology*, *45*(3), 699–733.

3 The Women of Piracy-Adjacent Trafficking and Their Victims

The human suffering and upheaval that accompanied the last decade of the twentieth century—particularly mass migration from the former Eastern bloc countries and nations undergoing widespread global socioeconomic upheaval spurred by austerity measures and politico-religious conflicts—brought widespread attention to human trafficking. Subsequent global campaigns to "end modern day slavery" have inspired the adoption of new state policies and the creation of new law enforcement mechanisms that include both private and public partnerships. Yet, attempts to clearly define human trafficking continue to evade scholars and policymakers alike. The United Nations provides a broad conceptualization of human trafficking as "the recruitment, transportation, transfer, harbouring or receipt of people through force, fraud or deception, with the aim of exploiting them for profit" (UNODC, 2022). The intentional broadness of the definition allows states to adopt and codify laws that address trafficking issues specific to their culture and context. Accordingly, human trafficking priorities have taken various forms globally, including sex trafficking, labor trafficking and exploitation, organ trafficking, child trafficking and exploitation, child soldering, and bride trafficking (Dragiewicz, 2014).

Despite this broad international legal conceptualization of human trafficking as encompassing a wide range of coerced labor, the traffic in women for the purposes of prostitution, or sex trafficking, dominates the popular cultural and policy realms which utilize sensationalized, inaccurate depictions of the sex industry and accounts of abuse within it to attract media attention (Weitzer, 2011, 2014; Dewey et al., 2018). Many scholars critique anti-trafficking interventions as little more than the product of these inaccurate depictions of the sex industry rather than reliable empirical evidence, resulting in a deeply flawed general rescue and rehabilitate approach that violates the rights of sex workers (Lerum & Brents, 2016). This chapter contributes an empirical example of ransom piracy in Somalia to highlight the interdependency between labor and

DOI: 10.4324/9781003225201-4

sex trafficking and piracy not yet explored in either literature. As we will see, these forms of trafficking are piracy adjacent, rather than central to the operations of piracy, yet are nonetheless important sources of revenue for pirates. Trafficking in Somalia, as elsewhere in the world, is a gendered phenomenon, with men and women generally filling unique roles as offenders and victims of forced prostitution, domestic servitude, bonded labor, forced labor, and slavery.

The Madams of Maritime Piracy: Sex Trafficking in Somalia

Fatuma approached the young woman who worked at a market stall in the center of Bosaso along Somalia's northern coastline. The young woman introduced herself as Warsame. She was beautiful, maybe 20 years of age, and exhibited a sense of naive confidence that Fatuma remembered having ages ago. Fatuma knew that Warsame, like almost all of her friends, dreamed of migrating to Europe and knew a journey like that wouldn't come easy. She told Warsame about a housemaid job with a family in Galkayo who had helped many Somali girls just like her to migrate to Italy. Warsame was giddy with excitement and could not believe her fortune. She thanked Faturma for the opportunity. The next morning, Warsame climbed into Fatuma's truck and they started off on their journey south.

The truck weaved through the streets of Garowe at nightfall before stopping alongside an unmarked compound. Fatuma stepped out and greeted, Mona, the woman who owned the brothel. Mona, like many women who own or manage sex industry venues, sold sex for many years. She and Fatuma met ten years prior when a downtrodden Fatuma begged for a job cleaning the facilities. They quickly developed a trusting relationship that evolved into a partnership of sorts. Mona inquired about the drive and the condition of the girl. The exchange was quick, as it always is. Fatuma struggled to recall how many girls she had recruited for Mona. Was it ten? Perhaps closer to 15 now? The demand for girls between the ages of 14 and 20 for sex with pirates exploded when piracy ransom money flooded into Garowe and other towns throughout Somalia. During her last transaction, she remembers Mona describing the pirates' appetites for women as insatiable because of the euphoria induced by regular khat use. Intoxicated and high on drugs, the pirates would pay for sex with two to three women per night. Fatuma started to wonder if she should just start selling the girls to the pirates directly to keep in their safe houses along the coast, bypassing Mona and the brothels altogether. As she gets back in her truck, she looks down at her phone and begins scrolling through her contacts to see whom she still knows in Eyl—the coastal hotspot of piracy.

The capture and enslavement of men and women is a well-known aspect of historical piracy. Pirates and privateers would openly purchase and sell captives at sea while sharing the waters with slave ships operating during the Atlantic and Indian slave trades. The Barbary pirates of North Africa would regularly purchase American sailors captured by Algerians, British fishermen kidnapped from their boats, and Spanish and Italian convicts, and force them to work on their ships or on land in their communities (Eltis, 1993; Ganser, 2020). Pirates operating in the Ancient Greek world reportedly regularly purchased and sold slave to the Greek elites (Garlan, 1987), and as early as the mid-seventeenth century, pirates kidnapped locals from Madagascar, at the time a popular pirate hideout, and sold them to ships (Hooper, 2011). Pirates have historically engaged in the kidnap, purchase, and sale of human beings, the behavior isn't new but framing it as "human trafficking" is. Throughout this chapter, I refer to these types of trafficking as "piracy-adjacent trafficking" because these forms of labor exploitation occur both in the context, and because, of piracy.

As is the case in many parts of the world, Somali women occupy a central role in the recruitment, transportation, and facilitation of women for sex and companionship. Women who work as madams, brothel owners, or in other intermediary roles in the sex industry worldwide typically have experience selling sex themselves and, as they gain experience and age, use their knowledge to recruit and manage younger women's sexual labor. Somali women are unique, however, in their simultaneous roles as sex industry intermediaries and resource purveyors for pirates. However, it is important to caution against assuming all women who sought out relationships with pirates were passively duped into participating in maritime piracy—albeit through sexual service. Researchers and journalists alike often dramatically refer to women being "lured" or "tricked" into the sex industry despite their knowledge that women sell sex to make ends meet, just as they might clean houses, care for other people's children, sell produce in the market, or engage in other forms of feminized labor that poor and working-class women routinely perform all over the world. Yet researchers and journalists never use verbs like "lured" and "tricked" to describe these other forms of labor that do not involve sex. These reductive descriptions fail to account for the agency of women who actively sought to enter into a relationship with individuals involved in a criminal enterprise. Cultural criminologists have advocated for a more gender conscious approach to analyzing women's involvement in transnational crime. Naegler and Salman's (2016) groundbreaking work on women's recruitment into the Islamic State of Iraq and the Levant suggests that women's sexual desires be taken seriously as potential motivation for criminal involvement. Just as men can be actively "seduced" into the thrill and excitement of engaging in terrorism, so too can women (Cottee & Hayward, 2011). Accordingly, not dismissing the possibility that women actively join pirate groups seeking lust, romance, companionship, and power presents the possibility of reimagining the women traffickers of piracy as facilitators whose work establishes culturally illicit but nonetheless mutually desired relations between women and pirates.

Interviews and discussion groups conducted with "key stakeholders" in Somalia, those who were deemed influential in their communities, as well as interviews conducted by other UNODC personnel (Yikona, 2013) in 2011 and 2012, reveal the experiences of victims of piracy-adjacent trafficking as detailed through the story of Fatuma and other victims. The workshops were held over five days in a hotel in Hargeisa, Somaliland. We met in a private conference room and participants were offered the opportunity to speak with me privately after the focus group sessions. Nearly all the women came to speak with me individually, where they felt they could speak more openly and honestly, and a handful of men followed up with me via email. Each session included about 10–15 participants who were chosen by local clan elders, regional government counter-piracy officials, and locally based UNODC interpreters. Some of the broader questions asked were, "How does piracy effect your community?", "Who is most negatively impacted by piracy?", "What would you like the international community to know about piracy"? The participants spoke with verdant passion, because the topic was very real and personal to them, their families, and their communities. At times, the volume in the room would get so loud it seemed like everyone was shouting at each other. My interpreter would try to keep up and assured me that Somalis talk loudly, and we weren't on the verge of a physical altercation. Men would stand up, hands and arms swinging wildly, to express the seriousness and urgency of their concerns.

One woman interviewed by UNODC reported being recruited from the Somaliland region of Somalia under similar circumstances as those discussed in Fatuma and Warsame's vignette. She and a friend were approached by a neighbor woman who offered them a job in the Puntland region and were told the money could help them achieve their dreams of traveling to the United Kingdom for an education. The young women envisioned they would work part time and attend university, something they had most likely heard about through diaspora friends or on social media. The girls left with the woman in the middle of the night without telling their parents and were taken to the coastal city of Bosaso where they were kept in a house with other young girls. The women were each assigned to a room of four to five pirates and forced to clean, wash clothes, and provide sexual services.

This testimonial, like others, reveals a more unique form of sexual labor than simply providing sexual gratification. Rather, these young women are providing sexual services as well as social reproductive support to pirates. They are more than sex slaves. Rather, they appear to be operating akin to "bush wives" held captive among West African militants. The bush wives of Sierra Leone's civil war (1991–2002) were kidnapped by rebel groups and forced to take part in the conflict as armed combatants, sex objects, mothers to children, and domestic servants (Coulter, 2011). Their experiences reveal the complexities of intersecting gender dynamics and sexual violence in the contexts of prolonged conflict and organized crime events (Marks, 2014).

Sexual labor is always a product of its cultural context. Somalia is a country guided and governed by the tenets of Islam and Somali customary law, *xeer*. The Somali Civil War of the 1990s triggered an ongoing social transformation that includes a rise in religious conservatism and an increasing regulation of Somali women's behavior and bodies (Abdi, 2007). A decrease in the education of girls and an increase in expectations of veiling are visible signs of this extreme social transformation (UNICEF, 2002; Abdi, 2007) With notions of honor (*sharaf*) and modesty (*xishood*) central to Somali culture and intertwined with a family's reputation, family members, usually older women, restrict younger female family members' interactions with unrelated males by limiting their mobility outside of the home, requiring a male family member accompany them in public, and relying on word of mouth of her whereabouts and behavior when supervision is not possible. A woman perceived as promiscuous, which could refer to as little as a young woman texting with a young man she is not related to or as much as a woman having sexual intercourse outside of wedlock, can bring shame onto her entire family and tarnish their name. She may no longer be eligible for marriage and her parents will have to support her until she dies. Her unwed siblings may also be looked upon unfavorably and have difficulty marrying.

Today, extramarital relations are forbidden in Somalia and the courtship process of arranging marriages is generally overseen by parents and clan elders. These social restrictions that limit contact between unmarried women and men have created an opportunity for women to serve as a conduit between pirates and unmarried (particularly young) women. Citizens of the Puntland region of Somalia described the rush of excitement in larger towns like Garowe and Bosaso when news spread of a successful piracy ransom payment for the release of hostages. Pirates would bring their sudden influx of money to these towns and spend it on cars, houses, drugs, and prostitution (World Bank, 2013). Some of the pirates are looking purely for sex, some are looking for women to satisfy their sexual desires and to carry out domestic activities with no intentions of marrying her, and a small number of pirates are looking for a wife and use the money to pay the dowry—the latter of which go through the traditional courtship rituals overseen by family. For those pirates looking for sex and/or a sexual companion who will also perform social reproduction tasks, they rely on women, typically older women, to facilitate the relations. These women tended to travel to more remote, impoverished rural areas to encourage young women to travel with them to a new destination where they will sell sex and/or act as pirates' companions. The recruitment process appeared to operate both across regions and even beyond national borders. The traffickers would travel to Somaliland and even as far as Ethiopia to recruit girls and transport them to popular pirate locales in the Puntland region (e.g., Bosaso, Galkayo, Garowe, Eyl). This created a geographical, cultural, and in the case of Ethiopia a linguistic distance between the young women and their families, making it more difficult for them to seek assistance and find opportunities for escape.

Many of the Somaliland citizens who were familiar with girls that were trafficked to Puntland for sex with pirates claimed the girls knew they were going to be having relations with pirates—seduced by the wealthy bad boy image of pirates at the time—but did so under the guise that they would marry the *pirates* and join them in their lavish lifestyle. During the peak years of piracy off the coast of Somalia, rumors were circulating throughout the country that women who married a pirate would earn a minimum of US$10,000–$20,000. Instead of paying the woman's family the dowry, the women believed rumors that they would be given the money. Faced with either having their families being paid by their suitor's family to arrange a marriage to a man who, due to lack of enforcement of domestic violence laws in Somalia, could beat her as he wished, pirates presented an opportunity for a marriage where domestic violence may also occur, but at least the relationship would be entered into by choice and the woman would receive thousands of dollars to do with as she wished. These young girls saw the money as a means out of their destitute poverty and as just the beginning of a life of luxury and the potential for travel to visit relatives in the diaspora all over the world, particularly London, Toronto, Seattle, and Minneapolis. Mothers of young women and well-respected elder women in the communities were desperate to counter these rumors with the realities that these girls would face as described by the handful of young women rescued from pirates by men in their community. One elder Somalilander man recalled how he and a few of his friends traveled to Garowe to rescue one of their community's young woman from pirates after her parents learned of her whereabouts. They described negotiating with the pirates and paying them a small amount of money to let the woman go with them; she had been beaten and was addicted to drugs.

Most young women sexually involved with pirates were not so fortunate, though. Many were simply thrown out on the streets when pirates tired of them or if they became pregnant. In these cases, the women experienced a double rejection of sorts, as society would cast them out due to their involvement in prostitution and drug use.

Women I interviewed also shared stories of married women who sought to leave their current husbands and marry pirate. These women saw marriage to a pirate as means to a better life and a way out of endless poverty and were not particularly interested in the fame and notoriety that comes along with marrying a pirate. The public-facing women associated with pirates, those the pirates allowed the world to see and not the ones beaten and drugged in the back room of the pirate safe house, were dressed in beautiful expensive clothes and sported the trendy new sunglasses imported from Europe and Dubai and rode around in the luxury cars not built for the dusty pothole ridden roads of Somalia. These beautiful, well-dressed women bolstered the pirates' image of success and bolstered recruitment into piracy for the young men who wanted to emulate their lifestyle and for those who knew they did not have the dowry funds to compete with pirates for a wife (Gilmer, 2017). In reality, many of the married women met the same fate as many of their unmarried, younger counterparts. Used and abused by pirates, the shame they brought upon their families meant they would be turned away if they tried to return to their husband and children. Consequently, they were forced to find a new way to exist and survive as a social outcast on the streets of Somalia's urban centers.

Careworkers of Maritime Piracy: Enslaved Hostages in Somalia

Leylo awoke with the rising sun and saw her father, a fisherman, smoking a cigarette as he prepared his net and boat along the water's edge. Her mother and seven younger siblings lay sound asleep next to her. She enjoyed being the first one awake and seeing her father off each morning. She watched as his feet sunk into the wet sand while he climbed into the boat and headed out to sea, the boat growing smaller and smaller until it disappeared among the waves of the Indian Ocean. Four days later his badly decomposed body washed ashore, his flesh torn and puckered with bullet holes, duct tape binding his hands and covering his mouth. Rumors spread like fire that he was picked up and murdered by the sailors aboard a nearby Russian warship. This rumor emanated from local fisherman who had seen the warship in the nearby vicinity in the past few days as well as a grainy video being shared on social media that showed what appeared to be several Somali men being shot on the deck of a Russian warship and their bodies thrown overboard into the sea.

In the aftermath of her father's death, Leylo and her family faced near starvation. The few relatives nearby had also fled the war years earlier and found themselves in a similar destitute situation and unable to assist them. When a friend approached their hut one morning with news that a man was asking around about hiring someone to wash clothes, they saw the opportunity as a lifeline. For next six months, every Sunday a man would bring a basket of clothes to their home and set it under the palm tree behind their house. They never saw his face, only his silhouette as he walked away. They would spend the next few days washing and ringing the clothes by hand and drape them in the trees to dry in the ocean breeze. When he returned to fetch the clean clothes, he left a pouch full of money in place of the basket. Whispers among the market eventually confirmed Leylo's suspicions that the man was part of the pirate group awaiting the ransom payment of two Norwegian owned-cargo ships anchored off the coastline. She shrugged the news off with indifference. The ocean may have brought pirates, but it also brought full bellies.

Care is not something typically associated with crime. In criminological and criminal justice scholarship, we might study the care provided to victims of crime or the ways in which agents of the state provide care to criminals as part of their confinement and rehabilitation, but we do not spend much time and energy thinking about care in the context of crime itself. The example of ransom kidnappings by Somalia pirates challenges the notion that carework and criminal actions are categorically distinct. Somali pirates, particularly women, consistently take actions to ensure the well-being of hostages—even if that the basic minimum threshold of well-being is keeping them alive (Wakeham & Gilmer, 2020). Likewise, as discussed at length in the section above, carework is central to the sex industry's everyday operations.

As ransom piracy attacks increased in frequency off the coast of Somalia around 2010, pirate groups quickly realized they could receive higher ransom payments for the safe return of hostages than the return of a ship and its cargo. Pirates' strategic goals in targeting ships accordingly shifted to identifying shipping vessels likely owned by larger shipping companies that would have some form of piracy and ransom insurance that could facilitate payment for the release of hostages (Shortland, 2017, 2019). The larger the ship, the more cargo and potential hostages it was likely to carry, the cleaner and more well kept the ship, the more likely the ship owner had some form of kidnap and ransom insurance. Consequently, traditional piracy tactics of attacking and plundering ships for their goods morphed into a lucrative kidnap and ransom business with human beings becoming the central commodity for (re)sale. Spatial strategies of violence and care became central to the secure captivation of hostages until a ransom payment was received (Gilmer & Dewey, 2022). Whereas pirates typically brought hostages from North America and Western Europe immediately to shore and moved throughout a network of camps and safe houses, the "typical" piracy hostages, low-income seafarers from the Global South remained aboard their hijacked ships anchored along the Somali coastline (Hurlburt, 2011).

Over a period of five years, Somali pirates held more than 3,600 hostages captive (ICC, 2022). Studies of hostage databases found that the length of time piracy hostages spent in captivity varied from months to years and depended greatly on the specific type of ship hijacked (De Groot et al., 2012). Those unfortunate enough to be captured aboard vessels with owners who abdicated their responsibilities to their crew found themselves captive for several years (Karimi, 2016). During captivity, many hostages were forced to continue carrying out duties associated with the daily maintenance and operation of the ship, such as electric and mechanical maintenance, fishing, cooking, and cleaning for all those aboard, including their captors (Gilmer & Wakeham, 2021). Pirates forced the crew to engage in this labor by threatening them with guns and violence, leading one former hostage to describe the conditions aboard his hijacked ship as "slavery". He describes crewmembers being beaten for not being able to fix a broken engine, others being forced to fish at gunpoint, all while the pirates sat and chewed *khat* and demanded hostages prepare food for them and bring them water at all hours (Moore, 2018).

Not all hijacked ships maintained the capacity for continued functioning as a self-sufficient community at sea. Gunfire and collisions with pirates' boats badly damaged some ships during the initial hijacking attack, making the crew—some of whom were themselves severely injured, sick, or incapable of caring for themselves—uncertain as to whether their ship could stay afloat. While higher-ranking pirates enlisted low-ranking male pirates to stay on board and monitor the hostages, even these low-ranking pirates had no desire to carry out the daily tasks of social reproduction viewed as "women's work" in Somalia. As the number of ships and hostages grew along Somali's coast, pirate groups recruited local women by to perform the carework associated with keeping hostages alive (Gilmer, 2019). The care was commodified in that the women were paid for their services, they were strangers to those they were caring for, and the work required time-limited interventions (England & Henry, 2013). Somalis participating in the UNODC workshop in Hargeisa, Somaliland, insisted that some women were going aboard the hijacked ships to perform tasks such as cooking and cleaning, while others worked from the shore preparing meals and cleaning laundry that would be delivered to the ships regularly. These arrangements varied by pirate group, depending on their unique circumstances. Whereas some pirate groups could rely on the hostages to perform carework for themselves *and* the pirates, most needed outside assistance and women stepped in to fill that need.

The employment of women to carry out the social reproduction of hostage-taking and slavery is not unique to ransom piracy. Free women were hired, and enslaved women were forced (but in some cases could earn small amounts of money) to care for slaves in the antebellum American South. Finley's (2020) groundbreaking research on women's involvement in what she labels the "intimate economy" of the slave trade details the various ways in which enslaved women worked sewing clothes, cooking food, washing and grooming bodies, monitoring slave jails, and performing housekeeping duties. She convincingly argues that women's crucial role in performing the day-to-day labor necessary to the functioning of the slave trade enabled the spread of slavery throughout the Southern United States and the expansion of profits associated with the sale and forced labor of human beings. The similarities between women caring for and preparing slaves for sale in the antebellum South and women caring for and preparing hostages for a ransom payment and release present an exciting new avenue of research. Both contexts demonstrate the complex interplay of gendered labor, social reproduction, and commodified bodies underscoring human trafficking.

Victims as Sexual Objects and Commodities

Race, ethnicity, citizenship, and gender play a central role in shaping piracy-adjacent trafficking in Somalia. Two distinct, yet intertwined, forms of forced labor revolve around pirates' hijacking ships for ransom payments: seafarers trafficked by pirates for ransom payments from their ship's owner, and women trafficked for sex with pirates. Pirates' receipt and redistribution of ransom payments links these otherwise distinct types of trafficking victims as pirates hold sailors hostage until they receive money from the ship owner, and then redistribute some of the ransom payment to purchase sex and companionship from women trafficked into prostitution. This redistribution of capital, whether it comes from ship owners' insurance companies, hostages' family members, or a government filtering it through another organization, eventually ends up in the hands of third parties profiting from trafficked women's forced sexual labor.

Women are both traffickers and victims of sex trafficking to pirates. Through their own words and the words of those who knew them, the women who did not intend to be involved in sexual relations with pirates described horrific stories of being locked in rooms and raped by groups of pirates, beaten and forced to perform domestic chores, and forced to take drugs to keep them from resisting or attempting to flee their circumstances. If the pirates grew tired of sex with the women, they would throw them out or abandon them as they moved onto a new location and found new women. These abuses and the treatment of the young women as sexual objects motivated many Somali elders, both men and women, to seek assistance from the regional Somali governments and international agencies. A "typical example" of recruitment, captivity and abuse unfolds with a pirate paying a trafficker to recruit and transport a girl to a pirate safe house, where he sexually assaults her with other pirates for days or months keeping her locked in the house, when they grow tired of her they tell her she is free to leave and return home. Some women are simply told to leave, whereas others are driven to the nearest town and told to get out of the car on a busy street corner of the pirates' choosing.

Somali women and women from neighboring countries are not the only women victimized by Somali pirates. Early twenty-first century, piracy off the coast of Somalia has functioned primarily as a ransom-based criminal enterprise involving the hijacking of commercial, fishing, or private vessels and holding of the seafarers and cargo hostage until a ransom is paid. While most of these hostages were poor male seafarers from developing countries, there have been several recorded incidents of kidnapping and ransoming white women. Several of these former hostages have written some of the few memoirs available that detail what it is like being held captive by Somali-based pirates (Chandler et al., 2012; Buchanan et al., 2013; Calitz and Hill, 2013), including a British woman sailing with her husband on their private yacht and a South African woman and her Italian partner working as crewmembers on board a yacht. In both cases, a ransom was paid to secure their release and rescue. A third woman, an American NGO worker, was kidnapped with her Danish male coworker, onshore in Somalia while they were being driven to the Galkayo airport. Whereas the first two women were released via ransom payment, the latter was rescued during a special force's military operation.

The women were kidnapped alongside male counterparts who also became victims of psychological torture and physical abuse throughout their captivity. Each woman recalled that they did not receive the same level of physical abuse as the men, but they felt under the constant threat of sexual assault. The American woman recounted being "roughed up somewhat" and constantly fending off unwanted touches, but the pirates who held her hostage never raped her. One can only speculate as to why their victimization experiences varied so immensely. Age may have played a factor. The elder British woman recounted the fewest instances of physical abuse, which may be attributed to Somalia's deep cultural respect for elders. Like historical piracy, marital status could also have been a protective factor. The two married women, the American and British hostages, were not raped. Both women were aware of the Islamic principles of not touching another man's wife and regularly reminded the pirates that they (as well as the pirates) were married.

The economics of the contemporary ransom piracy model may be what ultimately determined the difference in how Somali-based pirates treated African women, particularly Somali women versus high-profile white hostages. Whereas the rural Somali woman is recruited from a context of poverty and civil unrest, drugged, raped, beaten, and released into an unfamiliar landscape, the perceived potential ransom value of the British woman sailing on a private yacht is provided meals, doctors' visits and medication, and is spared from severe abuse. Both groups of captives are held hostage, but the Somali woman is disposable and cast out when she is no longer of use while the high-profile white hostages are an investment that can provide unimaginable financial security for their captors. This cost-benefit calculation can be seen in the initial ransom demands for the three high-profile female hostages: US$5 million for the British couple (McKenzie & Naik, 2010), US$10 million for the American woman and her Danish colleague (DeYoung & Jaffe, 2012), and US$10 million for the South African woman and her Italian partner (MacAskill et al., 2015). The families and friends of the hostages were unable to meet these wildly unrealistic demands, often based on the exaggerated wealth of Westerners portrayed in popular shows and Hollywood films, but the underlying assumption is that someone, somewhere will pay to have these women (and men) returned—unlike the sex trafficking victims whose families no longer deem them worthy of bringing home.

Piracy-Adjacent Trafficking in Gendered Global Context

Although the role of piracy-adjacent traffickers may be an emergent area of study, the phenomenon of women trafficking women is the norm worldwide. For example, nearly half of all trafficking victims in India self-reported their trafficker was a woman (Nair, 2004). The issue of women traffickers is now a well-established and growing area of study in transnational crime literature. The remaining sections refocus on the central role women play in facilitating forced prostitution in the context of Somali piracy.

Women in Transnational Trafficking

Scholars committed to understanding women who force other women into prostitution demonstrate that the role of trafficker is complex and diverse. Research by Siegel and de Blank (2010) identifies three ways women actively participate in the trafficking of other women as evidenced in Dutch court cases: supporters, partner in crime, and madams. Most women served in the capacity of "supporters"—someone who is subordinate to the leading trafficker but assists with various aspects of the crime such as recruiting and transporting victims. The second most common role of women traffickers in the Netherlands was women acting as "partners in crime". In these cases, women were usually in a relationship with the male trafficker and played an equal role as him in conducting various tasks. Madams, conversely, controlled trafficked women's sexual labor in sex industry venues they owned or managed.

Studies of women's participation in trafficking in the Balkans reveal a similar pattern as the Netherlands, with most women serving in the roles of partners in crime and supporters. Arsovska and Begum (2014) contend that women's peripheral participation in trafficking is a product of the Balkan's deeply entrenched patriarchal norms that require women to be subservient to and obey their husbands and fathers, and therefore, their roles in trafficking networks reflect their broader, passive role in society. Interestingly, however, a society's gender subordination is not always reflected in women's participation in trafficking. One can argue that Somali and Nigerian societies are very patriarchal in practice, however, yet women in both contexts play central leadership roles in the trafficking of other women (Noor, 2015; Amusan et al., 2017).

West African madams, particularly madams from Nigeria and Ghana, reportedly run some of the most intricate and expansive transnational trafficking networks in the world (Becucci, 2008; Campana, 2016). They play significant roles in trafficking, including organizing, recruiting, transporting, exploiting, and enforcement. Although not all madams play a central role in trafficking networks that facilitate women's migration (Mancuso, 2014), scholars claim that gender, class, and women's acceptance in positions of power in the public sphere may contribute to increased opportunities for some West African women to take on leadership roles in transnational human trafficking operations. Siegel (2014) argues that contrary to women who rise to power in the Italian mafia upon the death or imprisonment of their husbands, Nigerian and Ghanian madams took the initiative themselves to establish, head, and coordinate the various tasks of the human trafficking business, including recruitment of girls; facilitating transportation; coordinating housing, food, and clothing for the sex workers; and recruitment of clients.

Madams and their trafficking victims share the desire for geographical and social mobility (Lo Iacono, 2014). The emergence of women-led, large-scale trafficking organizations in their native countries and the ability to make quick money in Europe helped establish a sizable network of opportunity and exploitation. Researchers and practitioners have begun engaging with complexities of this network to rethink transnational Nigerian trafficking as indentured sex work migration to better capture the voluntary entry and negotiated levels of exploitation that occur prior to migration (Rizzotti,

2022). Doing so opens additional entry points for understanding and addressing the women-led trafficking phenomenon beyond the perpetrator/victim dichotomy and illuminates the nuances of power, culture, agency, and migration at work in trafficking networks.

Research conducted in China and Cambodia presents yet another unique way women are involved in trafficking, including as child traffickers for illegal adoption. Data collected from interviews with incarcerated traffickers reveal that internal, domestic child trafficking is opportunistic, by chance, once-off, and short-term strategy as opposed to the more highly structured trafficking networks described in other studies of women supporters, partners in crime, and madams (Shen, 2016). The women reported obtaining the babies through various channels, including accepting unwanted babies, picking up abandoned babies, and kidnapping with minimal initial investment or need for violence (Steffensmeier, 1983; Shen, 2016). Despite the lack of sophisticated organized networks, all the women interviewed reported relying on one or two people, typically a spouse, friend, or coworker, to carry out some of the trafficking tasks such as looking for a buyer, transportation, and sheltering (Shen et al., 2013; Shen, 2016).

Female counterparts involved in child trafficking in Cambodia also appeared to operate outside of a structured trafficking network. Drawing from prison and police records and interviews with alleged offenders, Keo et al. (2014) discovered that 80 percent of incarcerated traffickers in Cambodia are uneducated women from low socioeconomic backgrounds. Many of the female traffickers had previously been destitute and worked in commercial sex work—which, in the recent global war on trafficking, has retroactively earned them the label prior trafficking victims (Keo, 2010). Accordingly, Keo et al. (2014) contend the phenomenon of women trafficking women and children in Cambodia is a grossly exaggerated, low profit, unsophisticated opportunistic crime exacerbated by a lack of legitimate employment opportunities. Somali women who traffic women in Somalia are also operating in a bleak economic environment with few legitimate employment opportunities and the women they recruit come from a similar low socioeconomic background with little to no education.

Theorizing the Madams and Careworkers of Maritime Piracy

Women traffickers associated with Somali piracy share motivations for offending and operational characteristics with women traffickers in other geographical locations. Like the child traffickers of China and Cambodia, the Somali women account for the majority, if not all, of the individuals trafficking women for sex with pirates. As previously discussed, the gendering of the trafficking role in Somalia is most likely a product of the strict cultural norms that forbid unmarried women from mingling with and courting unmarried men without adult supervision. The traffickers, then, act as the intermediary between the women and the pirates; an older woman talking with a younger woman is not an act that immediately raises red flags.

The Somali traffickers also resemble their Chinese and Cambodian counterparts in that they, too, seem to be uneducated, poor, and looking for economic opportunities. Somalia has one of the world's lowest adult literacy rates, with an estimated 14 percent literacy rate among Somali women, forcing many to rely on the informal economy, illicit activities, and entrepreneurial ingenuity to survive (Williams & Cummings, 2017). Lastly, although the trafficking networks stretched across Somali regions and even into Ethiopia, testimonials from former victims did not identify other perpetrators beyond the women that recruited them. It is conceivable that the women traffickers of Somalia keep their networks small and only rely on one or two closely trusted people to carry out tasks as needed. As is the case worldwide, those who meet the international legal definition of "trafficker" usually share many life circumstances and characteristics with those who meet the definition of "trafficking victim".

The traffickers of Somali piracy appear to share the most commonalities with the madams of West Africa. Despite operating in very religious and patriarchal societies, the complex histories of countries like Somalia and Nigeria that include internal conflict and war have enabled women to gain power and entry into criminal enterprise through their own initiative and entrepreneurial skills. Just as the Yoruba women of Nigeria have historically translated their economic and social roles into power and influence, the women of Somalia stepped in to keep markets and family businesses running during and after Somalia's civil war (Denzer, 1994).

The desire for geographical and social mobility motivates both Somalia and West African traffickers and their victims. The traffickers exploit the young women's desire to travel abroad and improve their livelihood circumstances. Whereas the madams of Nigeria transport their recruits transnationally to Europe where they are exploited, the Somali women indulge the recruits' desires to move abroad for a better life by claiming they can quickly earn money to travel to Europe. Somali and Nigeria boast large diaspora communities. Stories from expats living in Europe and North America trickle back to their home countries and inspire the youth to move abroad to seek more financial opportunities and security. This creates a situation ripe for exploitation. Traffickers are not the only ones capitalizing on the youth's desire for mobility in these contexts. The extremist group al-Shabaab has also been known to promise Somali youth an education in Europe as a recruitment tool (Omenma et al., 2020).

It is difficult to attempt to situate what little is known about the women traffickers of Somali piracy into the broader context of women-led trafficking enterprises across the globe. Perhaps, this chapter is just the beginning of unraveling what may be a bigger geo-historical phenomenon of the madams of maritime piracy. The importance of beginning to engage with this unique women-led role of piracy cannot be overstated. Despite its many valid critiques (see Galusca, 2012; Dewey et al., 2020; Mai et al., 2021), international law and policy directed at human trafficking still focuses on inaccurate and misguided representations of sex trafficking and ignores the much more prevalent problem of labor trafficking. As scholars take more seriously the intersections between sex trafficking, war, and other transnational crimes (Okolie-Osemene & Okolie-Osemene, 2019; Petrich & Donnelly, 2019; Njoku, et al. 2022), it is imperative to also consider the crime of maritime piracy. Studies of contemporary piracy must engage more fully with the role of women as both potential perpetrators and victims of sex trafficking as it relates to piracy. There are also fruitful avenues for revisiting historical piracy archives to rethink the relationship between the forced movement and exploitation of human beings associated with maritime piracy.

References

Abdi, C. M. (2007). Convergence of civil war and the religious right: Reimagining Somali women. *Signs: Journal of Women in Culture and Society*, *33*(1), 183–207.

Amusan, L., Saka, L., & Ahmed, Y. B. (2017). Patriarchy, religion and women's political participation in Kwara state, Nigeria. *Gender and Behaviour*, *15*(1), 8442–8461.

Arsovska, J., & Begum, P. (2014). From West Africa to the Balkans: Exploring women's roles in transnational organized crime. *Trends in Organized Crime*, *17*(1), 89–109.

Becucci, S. (2008). New players in an old game: The sex market in Italy. In D. Siegel & H. Nelen (Eds.), *Organized crime: Culture, markets and policies* (pp. 57–69). New York: Springer.

Buchanan, J., Landemalm, E., & Flacco, A. (2013). *Impossible odds: The kidnapping of Jessica Buchanan and her dramatic rescue by SEAL team six.* New York: Simon and Schuster.

Calitz, D., & Hill, U. (2013). *20 months in hostage hell.* New York: Penguin Global.

Campana, P. (2016). The structure of human trafficking: Lifting the bonnet on a Nigerian transnational network. *British Journal of Criminology*, *56*(1), 68–86.

Chandler, P., Chandler, R., & Edworthy, S. (2012). *Hostage: A year at gunpoint with Somali pirates*. Chicago: Chicago Review Press.

Cottee, S., & Hayward, K. (2011). Terrorist (e) motives: The existential attractions of terrorism. *Studies in Conflict and Terrorism*, *34*(12), 963–986.

Coulter, C. (2011). *Bush wives and girl soldiers*. Ithaca: Cornell University Press.

Denzer, L. (1994). Yoruba women: A historiographical study. *International Journal of African Historical Studies*, *27*(1), 1–39.

De Groot, O. J., Rablen, M. D., & Shortland, A. (2012). Barrgh-gaining with Somali pirates (No. 74). Economics of Security Working Paper.

Dewey, S., Crowhurst, I., & Izugbara, C. (2018). Globally circulating discourses on the sex industry: A focus on three world regions. In S. Dewey, I. Crowhurst, C. Izugbara (Eds.), *Routledge international handbook of sex industry research* (pp. 186–197). Oxfordshire: Routledge.

Dewey, S., Crowhurst, I., Zheng, T., & Blanchette, T. (2020). Control creep and the multiple exclusions faced by women in low-autonomy sex industry sectors. *Vibrant: Virtual Brazilian Anthropology*, *17*, 1–25.

DeYoung, K., & Jaffe, G. (2012, January 25). Navy SEALs rescue kidnapped aid workers Jessica Buchanan and Poul Hagen Thisted in Somalia. Washingtonpost .com. Retrieved from https://www.washingtonpost.com/world/national-security /us-forces-rescue-kidnapped-aid-workers-jessica-buchanan-and-poul-hagen -thisted-in-somalia/2012/01/25/gIQA7WopPQ_story.html

Dragiewicz, M. (Ed.). (2014). *Global human trafficking: Critical issues and contexts*. Oxfordshire: Routledge.

Eltis, D. (1993). Europeans and the rise and fall of African slavery in the Americas: An interpretation. *American Historical Review*, *98*(5), 1399–1423.

England, K., & Henry, C. (2013). Care work, migration and citizenship: International nurses in the UK. *Social and Cultural Geography*, *14*(5), 558–574.

Finley, A. J. (2020). *An intimate economy: Enslaved women, work, and America's domestic slave trade*. Chapel Hill: UNC Press Books.

Galusca, R. (2012). Slave hunters, brothel busters, and feminist interventions: Investigative journalists as anti-sex-trafficking humanitarians. *Feminist Formations*, *24*(2) 1–24.

Ganser, A. (2020). *Crisis and legitimacy in Atlantic American narratives of piracy*. Cham: Palgrave Macmillan.

Garlan, Y. (1987). War, piracy and slavery in the Greek world. *Slavery and Abolition*, *8*(1), 7–21.

Gilmer, B. (2017). Hedonists and husbands: Piracy narratives, gender demands, and local political economic realities in Somalia. *Third World Quarterly*, *38*(6), 1366–1380.

Gilmer, B. (2019). Invisible pirates: Women and the gendered roles of Somali piracy. *Feminist Criminology*, *14*(3), 371–388.

Gilmer, B. V., & Dewey, S. C. (2022). Captive calculations and benevolent abandonment: Ransom piracy and the carceral occupation of hostage ships along the coast of Somalia. *Political Geography*, *95*, 102586.

Gilmer, B. V., & Wakeham, J. (2021). Between ransom and release: Exploring Caringscapes of ransom kidnappings by Somali pirates. *Journal of the Middle East and Africa*, *12*(3), 301–319.

Hooper, J. (2011). Pirates and kings: Power on the shores of early modern Madagascar and the Indian Ocean. *Journal of World History*, *22*(2), 215–242.

Hurlburt, K. (2011). The human cost of Somali piracy–oceans beyond piracy (one earth future foundation. Retrieved from http://oceansbeyondpiracy.org/sites/default/files/human_cost_of_somali_piracy.pdf

(ICC) International Chamber of Commerce Commercial Crime Services. (2022). Somali piracy: Last three hostages freed but threat still exists. Retrieved from https://www.icc-ccs.org/index.php/1295-somali-piracy-last-three-hostages-freed-but-threat-still-exists

Karimi, F. (2016, October 23). Somali pirates free 26 hostages after nearly 5 years in captivity, group says. *CNN*. Retrieved from https://www.cnn.com/2016/10/23/africa/somalia-pirates-release-hostages/index.html

Keo, C. (2010). *NGO joint statistics project: Database report on trafficking and rape in Cambodia 2009*. Phnom Penh: eCPAT-Cambodia.

Keo, C., Bouhours, T., Broadhurst, R., & Bouhours, B. (2014). Human trafficking and moral panic in Cambodia. *Annals of the American Academy of Political and Social Science*, *653*(1), 202–224.

Lerum, K., & Brents, B. G. (2016). Sociological perspectives on sex work and human trafficking. *Sociological Perspectives*, *59*(1), 17–26.

Lo Iacono, E. (2014). Victims, sex workers and perpetrators: Gray areas in the trafficking of Nigerian women. *Trends in Organized Crime*, *17*(1), 110–128.

MacAskill, E., Milne, S., & Swisher, C. (2015, October 8). Italian intelligence lied about hostage rescue to hide ransom payment. Theguardian.com. Retrieved from https://www.theguardian.com/world/2015/oct/08/italian-intelligence-lied-hostage-rescue-bruno-pelizzari-debbie-calitz

Mai, N., Macioti, P. G., Bennachie, C., Fehrenbacher, A. E., Giametta, C., Hoefinger, H., & Musto, J. (2021). Migration, sex work and trafficking: The racialized bordering politics of sexual humanitarianism. *Ethnic and Racial Studies*, *44*(9), 1607–1628.

Mancuso, M. (2014). Not all madams have a central role: Analysis of a Nigerian sex trafficking network. *Trends in Organized Crime*, *17*(1), 66–88.

Marks, Z. (2014). Sexual violence in Sierra Leone's civil war: 'Virgination', rape, and marriage. *African Affairs*, *113*(450), 67–87.

McKenzie, D., & Naik, B. (2010, November 14). British couple freed, held more than a year by Somali pirates. Cnn.com. Retrieved from https://www.cnn.com/2010/WORLD/africa/11/14/somalia.couple.released/index.html

Moore, M. S. (2018). *The desert and the sea: 977 days captive on the Somali pirate coast*. Harper Wave.

Naegler, L., & Salman, S. (2016). Cultural criminology and gender consciousness: Moving feminist theory from margin to center. *Feminist Criminology*, *11*(4), 354–374.

Nair, P. M. (2004). *A report on trafficking in women and children in India, 2002–2003*. New Delhi: National Human Rights Commission.

Njoku, E. T., Akintayo, J., & Mohammed, I. (2022). Sex trafficking and sex-for-food/money: Terrorism and conflict-related sexual violence against men in the Lake Chad region. *Conflict, Security and Development*, *22*(1), 79–95.

Noor Mohammed, H. (2015). Kenya and Somalia: Fragile constitutional gains for women and the threat of patriarchy. *African Security Review*, *24*(4), 458–474.

Okolie-Osemene, J., & Okolie-Osemene, R. I. (2019). Nigerian women and the trends of kidnapping in the era of Boko Haram insurgency: Patterns and evolution. *Small Wars and Insurgencies*, *30*(6–7), 1151–1168.

Omenma, J. T., Hendricks, C., & Ajaebili, N. C. (2020). al-Shabaab and Boko Haram: Recruitment strategies. *Peace and Conflict Studies*, *27*(1), 2.

Petrich, K., & Donnelly, P. (2019). Worth many sins: Al-Shabaab's shifting relationship with Kenyan women. *Small Wars and Insurgencies*, *30*(6–7), 1169–1192.

Rizzotti, M. (2022). Chasing geographical and social mobility: The motivations of Nigerian madams to enter indentured relationships. *Anti-Trafficking Review*, *18*, 49–66.

Shen, A. (2016). Female perpetrators in internal child trafficking in China: An empirical study. *Journal of Human Trafficking*, *2*(1), 63–77.

Shen, A., Antonopoulos, G. A., & Papanicolaou, G. (2013). China's stolen children: Internal child trafficking in the People's Republic of China. *Trends in Organized Crime*, *16*(1), 31–48.

Shortland, A. (2017). Governing kidnap for ransom: Lloyd's as a "private regime". *Governance*, *30*(2), 283–299.

Shortland, A. (2019). *Kidnap: Inside the ransom business*. Oxford: Oxford University Press.

Siegel, D. (2014). Women in transnational organized crime. *Trends in Organized Crime*, *17*(1/2), 52–65. https://doi.org/10.1007/s12117-013-9206-4

Siegel, D., & De Blank, S. (2010). Women who traffic women: The role of women in human trafficking networks–Dutch cases. *Global Crime*, *11*(4), 436–447.

Steffensmeier, D. J. (1983). Organization properties and sex-segregation in the underworld: Building a sociological theory of sex differences in crime. *Social Forces*, *61*(4), 1010–1032.

UNICEF (United Nations Children's Fund). 2002. Survey of primary schools in Somalia, 2001/2. Technical Report, Vol. 1. UNICEF Somalia Support Center.

UNODC (United Nations Office on Drugs and Crime). (2022). Human trafficking. Retrieved from https://www.unodc.org/unodc/en/human-trafficking/human-trafficking.html

Wakeham, J., & Gilmer, B. V. (2020). Care in uncaring places: Exploring Somali piracy hostage incidents. *British Journal of Criminology*, *60*(5), 1242–1259.

Weitzer, R. (2011). Sex trafficking and the sex industry: The need for evidence-based theory and legislation. *Journal of Criminal Law and Criminology*, *101*, 1337–1370.

Weitzer, R. (2014). New directions in research on human trafficking. *Annals of the American Academy of Political and Social Science*, *653*(1), 6–24.

Williams, J. H., & Cummings, W. C. (2017). Education from the bottom up: UNICEF's education programme in Somalia. In J. Williams & W. Cummings (Eds.), *Development assistance for peacebuilding* (pp. 137–152). Oxfordshire: Routledge.

World Bank. (2013). *Pirate trails: Tracking the illicit financial flows from pirate activities off the Horn of Africa.* Retrieved from https://alabama.app.box.com/integrations/officeonline/openOfficeOnline?fileId=996218771772&sharedAccessCode=

Yikona, S. (2013). *Pirate trails: Tracking the illicit financial flows from pirate activities off the Horn of Africa.* Washington D.C.: World Bank Publications.

4 Policing, Punishment, and Counter-Piracy-Involved Women

Historically, punishments for piracy were contingent on where the crime was committed and prosecuted and the citizenship, age, and race of the perpetrators and victims. Gender also influenced the method and severity of punishment for piracy. Social constructions of criminal women, including pirates, have changed over time to reflect broader trends in criminological thought, and what constitutes a criminal woman, and more specifically a *punishable* woman, is shaped by academic discourse and institutional agendas (Snider, 2003), in tandem with popular understandings of crime and punishment. Feminist criminologists have long argued that the male-dominated fields of policing and corrections, as well as the male-dominated discipline of criminology, place men in the position of the powerful knower with the unearned privilege to frame *what* we know about and *how* we continue to engage with women offenders (Chesney-Lind & Morash, 2013; Cook, 2016).

For at least the past two decades, academics have noted a shift from the general underrepresentation of women in crime to an overrepresentation of women in research and prosecution of gender-specific crimes such as prostitution, abortion, and infanticide (DeKeseredy, 2000). This, along with an increasing use of imprisonment in response to women's behavior and the criminalization of petty offenses, is correlated with 17 percent increase over the past decade in the population of incarcerated women (Penal Reform, 2021). In the United States alone, that number has increased fivefold since the 1980s, and women in US prisons now account for 30 percent of the world's total number of incarcerated women (Sentencing Project, 2022; Kajstura, 2018). Patriarchal governments design and enforce laws and use imprisonment to reinforce traditional gender and cultural norms for women. For example, Sierra Leone began enforcing loitering laws that prohibited women from being outdoors at night under the guise of discouraging "sexual promiscuity" and laws prohibiting "witchcraft" disproportionately

DOI: 10.4324/9781003225201-5

result in women's incarceration in the Central African Republic (Penal Reform, 2021).

While the unequal enforcement of some laws certainly victimizes women, girls and women who commit violent crimes remain largely understudied (Pasko and Chesney-Lind, 2018). Rather than suggesting women have become more violent over time, as many popular cultural commentators began to note in the early twenty-first century, Irwin et al. (2018) suggest that the justice system is finally taking girls' and women's violence more seriously. Violence is always contextual and gendered. For example, violence committed by boys or men, particularly in wartime, is valorized as a display of patriotism and agency, whereas violence committed by girls or women is generally interpreted as a failure to "protect" them from the dangers of society or a flaw in their femininity (Carrington et al., 2010; Sjoberg & Gentry, 2007). Accordingly, prevailing criminal justice approaches envision female offenders as having character flaws and problematic personality traits characterized by the courts as manipulative, hysterical, wildly sexual, untrustworthy, and immoral (Gaarder & Belknap, 2002; Lopez & Pasko, 2021; Mallicoat, 2007; Pasko, 2010; Pasko & Lopez, 2018; Schaffner, 2006).

Laws and police tactics are established and enforced in ways that reduce women to their bodies. This is a long-standing phenomenon experienced worldwide by women and most notable in relation to sexuality, reproduction, and abortion (Quinlan, 2017). As women are reduced to little more than a receptacle for a uterus, they continue to contend with state policies and practices that on one hand are aimed at shaping what they are permitted to do with their bodies, and on the other hand are ineffectively (or unintentionally) enforced when women's bodies become sites of violence (Atkins & Hale, 2018; Collins & Dunn, 2018; Roychowdhury, 2020). How women are reduced to sexualized, racialized bodies and wombs through law and criminal justice practices remains understudied (Ritchie & Jones-Brown, 2017). As Chesney-Lind (2017) suggests, "No complete theory of gender and the state will be possible until we fully appreciate and document the role of police, courts, and prisons in the ratification and enforcement of male power over women's sexualized bodies" (1). The knowledge–power nexus of criminal justice is a key space for engaging with these matters.

Knowledge claims about women who offend shape how criminal justice systems punish women offenders. As understandings about women who offend change over time, so do understandings about what constitutes acceptable and effective punishments for their transgressions. Snider (2003) argues that these knowledge claims, including those produced by feminist criminologists, have contributed to the "incarceration spiral" of recent times. From the eighteenth century to roughly the 1970s, female offending was understood through a lens of law, science, and religion. Accordingly, women offenders were perceived morally reprehensible and either deserved severe punishment for her deviance from social expectations, or as in the case of many areas of the United States, needed reforming through religious guidance, supervision, and discipline.

Modern-day understandings of female offending continue to use male offending as the norm and benchmark for understanding women's participation in crime (Daly & Chesney-Lind, 1988; Smart, 1995; Balfour & Comack, 2021). Feminist criminologists' call to study the growing number of women who became offenders and jail and prison inmates produced theories such as the chivalry hypothesis and offender-as-victim hypothesis. The chivalry hypothesis suggested that data proved that women were being treated more leniently than their male counterparts (Morris, 1987; Spohn et al., 1987; Smart, 1995). The offender-as-victim hypothesis argued that most women offenders were victims themselves, which contributed to their propensity to offend (Chesney-Lind, 2002). Both well-intentioned scholarly interventions aimed at making women more visible in criminology had material consequences for women offenders. Whereas leniency arguments were used to punish women more equally to their male counterparts (rather than to punish less all around), the offender-as-victim argument helped constitute and reinforce the "resilient, resistant female offender" that prison reformers, social workers, psychologists sought to heal and transform through gender-responsive approaches focused on women offenders' often extensive trauma histories (van Wormer & Bartollas, 2021; Widom & Osborn, 2021). These academic constructions of the woman offender were not just *heard* (Snider, 2003) but were *mobilized* by policymakers and practitioners.

This chapter explores how various approaches and techniques for punishing pirates—from hangings to jail sentences to amnesty—have been adapted or abandoned when bringing female pirates to justice. It also examines the instrumentalization of gender and motherhood by contemporary counter-piracy programming efforts to suppress piracy off the coast of Somalia. In so doing, I argue that punishment for piracy is a phenomenon as gendered as piracy itself.

Policing and Punishing Women Pirates

The two women stood defiantly side by side with their heads raised and hands shackled behind them. Ten days prior, their male counterparts stood in the same courtroom and were convicted of piracy, forcibly dragged outside to the gallows, and hung by the neck until dead. Captain Jack Rackham's decaying body was still nearby on display, locked up in an iron gibbet as a warning to others. Witnesses took turns sharing testimony of the women's cruelty and violence at sea, describing in painful detail their unacceptably masculine demeanor, and seeming ambivalence to murder. As the sun set outside of the High Admiralty court in Spanish Town, Jamaica, the women knew they had little chance of escaping the same fate as their fellow crew. They offered no statement in their own defense as the judge solemnly pronounced Anne Bonny and Mary Read guilty and sentenced them to immediate death by hanging.

The judge rose from his chair with a muffled thump and directed the court officer to escort the women out to the gallows. As the officer nodded, the women recoiled and turned to the judge shouting that they were both pregnant. For a moment the courtroom was silent as all who were present stood motionless and stunned. The judge's face twisted in anger and disappointment. It was not customary to sentence a pregnant woman to death, let alone two. Left with no choice, he turned and addressed the courtroom. There would be a stay of execution until the children were born so as not to take an innocent life along with that of a guilty one. Upon giving birth, the women were to be hanged. The court officer grasped each woman's arm with his hand and quickly led them out of the room back to their cells where only one would eventually leave alive.

The trial of Anne Bonny and Mary Read is arguably the most well-known historical example of women standing trial for piracy. Although the outcome of the trial is undisputed, and both women were pregnant during the hearing, details of each woman's fate are hazy at best. Whereas Read eventually died in prison, purportedly from illness associated with childbirth, Bonny's fate remains a mystery with some speculating that she was eventually set free with her infant child (Rediker, 1996; Hernandez, 2009; Cordingly, 2006). In any case, neither woman was executed for piracy, and their trial marked the beginning of a longer historical trend toward lenient sanctions for women pirates. Those found guilty tended to be granted amnesty, acquitted, or handed down a lighter punishment than initially charged.

The complex legal history of piracy takes place against a backdrop of shifting state policies, elastic jurisdiction, and contested sovereignties. Indeed, ambiguities still exist about when piracy first became a crime of universal jurisdiction (Benton, 2011). Flexible strategies adopted throughout history to bring pirates to justice include either trying them in municipal/metropolitan courts, extending sovereign reach and establishing remote courts in foreign territories, or a combination thereof. However, uncovering and unpacking the policing and punishment of women in relation to piracy-related crimes is a difficult task. The sparse historical records that exist tend to be incomplete, and there appears to be a hesitancy to arrest women and bring them before courts for piracy in modern times; rather, they are dealt with privately by families which may include a complete restriction of the woman's movement outside of the home or having her relocate to live with relatives elsewhere where her crimes are less likely to be known by community members. This chapter's remaining sections explore examples of policing and punishing women pirates in different historical and geographical contexts.

England and the Americas

Piracy trials held in English courts spanning the 1500s–1700s paint a picture of women heavily involved in shore-based piracy networks. The Tudor regime was unable to quell these shore-based activities despite proclaiming that anyone aiding and abetting pirates, particularly those who were receiving and selling goods on their behalf, would face the same punishment as pirates—death (Appleby, 2013). A sampling of court convictions from piracy trials in England shows that a relatively large number of defendants were acquitted and the public executions for piracy seemed to be reserved for those prosecuted for attacking ships at sea (Gibbs, 2019).

Women widely participated in shore-based piracy as receivers of goods through localized and opportunistic activities in which they took advantage of gendered expectations and their knowledge of local trade networks. Trial transcripts highlight how some women involved in piracy used their marriages to admiralty officers and sheriffs to effectively operate with impunity for extended periods of time (Appleby, 2013). Other well-connected receivers included pirates' mothers and spouses, such as pirate John Piers' mother Anne Piers, who was arrested and interrogated in 1581 on charges of receiving stolen goods and witchcraft. It was believed that Piers' abilities as a witch enabled her to successfully receive and sell goods without being caught. After interviewing several witnesses, she was found to be a woman of "loose life" but acquitted of the witchcraft charges. There are conflicting accounts of whether she was found guilty or acquitted for the handling of stolen goods, but it was generally agreed that she was never imprisoned for her actions (Appleby, 2016; National Archives, 2022).

Examples of well-connected English piracy are rare, as most women caught receiving goods were too poor to pay court fines and were released with little to no punishment. Dealing with women receivers was a difficult task for the court and investigating their crimes and producing witnesses proved complicated. In many cases, the women pleaded ignorance of both the goods they were moving and the persons who asked them to move the goods. Accordingly, the court was left with few choices but to deem their participation "unintended" and let them off with a small fine or an acquittal. Compared to the pirating of vessels, receivership was viewed as an economy of the poor, and the cases seemed trivial to an already overburdened court. Women's participation in receiving, bartering, and exchanging pirated goods, although the transactions may have appeared small scale, was part of a much larger, complex network that provided huge profits for all participants, and particularly for pirates (Appleby, 2013).

Eighteenth-century America witnessed several women on trial for activities related to piracy. Two such women were Mary Harley (Harvey) and Mary Critchett (Crickett). Mary Harvey was pirating with her husband and a group of men up off the coast of Virginia and, when caught, all were sentenced to death for crimes of piracy. However, whereas the men were hung, Harley was released. Years later, Critchett was brought from England to Virginia as a convict along with others to labor off her debt to society. She and some of the other prisoners managed to escape and she joined with a group of pirates who successfully attacked a vessel but were later apprehended in the Chesapeake Bay. Critchett did not receive the same mercy as Harley when she and her fellow pirates were tried, convicted, and hung for piracy in Williamsburg, Virginia (Rediker, 2004; Creighton & Norling, 1996; National Park Service, 2020).

Another eighteenth-century American woman pirate, Rachel Wall, was born in Pennsylvania about 1760 and moved to Boston after marrying her husband. Her husband was often in trouble with the law for petty crimes and Wall actively helped him with his criminal endeavors and even helped him escape from jail on one occasion. They soon rounded up some accomplices and took their criminal activities to the ocean to get rich with a scheme that had Hall feign distress to nearby ships. The sailors of the other vessels would come to her aid, board the ship, and be killed by Hall's husband and the other pirates. After the murders, Hall's crew would plunder the sailors' ship and sell the bounty back on shore. Hall's group pirated off the coast of New England for approximately one year until her husband and other crew members were killed at sea during a storm. She and the other surviving pirates were rescued and brought back to Boston by other seafarers where she would continue pirating from the shore by looting ships anchored in port. Eventually, she was caught, pled guilty, paid a fine, and was released. Hall was repeatedly arrested for similar crimes and when she could not pay the fines, she served three years of indentured labor. She was tried and found guilty of robbing a woman and hanged on Boston Common in 1879, making her the last person to be hanged in the state of Massachusetts (Vargo, 2015).

China

Pirates menaced the South China seas during the eighteenth and nineteenth centuries. The imperial government was continually modifying their response to pirates to include military, pacification, and judicial approaches (Antony, 1992, 2012). The most direct approach was the use of military campaigns aimed at annihilating pirates. These campaigns were unsuccessful due to the imperial forces being inferior to the enormous fleet of ships amassed by the pirates. The government also attempted pacification strategies that would include offering pirate leaders the opportunity to surrender in return for monetary rewards and being appointed leadership positions or titles in the imperial forces. While some pirates would take up these offers, most did not want to be relegated to the orders of the imperial forces scoffed at the monetary rewards in comparison to the money they were making through their illicit endeavors.

Two notable Chinese women, Ching Shih and Ng Akew, avoided punishment by accepting pacification agreements, also referred to as amnesty agreements, offered by the imperial government that ensured they would not be prosecuted if they surrendered and agreed to stop pirating. Despite the ferociousness and size of Shih's fleet, scholars suggest that she accepted the amnesty offer due to increasing tensions among the leadership of her piracy confederation (MacKay, 2013). She used the opportunity to safely exit piracy by surrendering in 1810 and negotiating amnesty for her and her husband, the continued possession of several vessels, and secured military commissions for her husband and several fellow pirates. Effectively, Shih provided upward mobility for many of the men around her (Murray, 1981). Akew, on the other hand, was a skilled entrepreneur involved in opium smuggling and business dealings with pirate groups and other powerful foreign and Chinese partners. Her connections and relationships with elites ultimately helped her achieve amnesty in lieu of punishment for associating with pirates (Kwan, 2020).

Pirates who chose to forego pacification efforts found themselves at the mercy of the imperial courts, whose judicial approach dealt pirates some of the harshest penalties afforded to criminals. Under the Qing government, pirates found guilty were normally decapitated, with pirates' severed heads transported to coastal communities and displayed on posts as a warning to other pirates and pirate supporters. The alternative to decapitation was death by slicing, known as *lingchi* (Antony, 2012). *Lingchi* is an ancient Chinese form of capital punishment that involved tying the convicted person to a wooden post while the executioner administering cuts to the bare flesh, slowly moving from one area of the body to another, until the person bled to death in excruciating pain followed by bodily dismemberment and placement of the pirates' disarticulated body parts inside of a basket. The practice was eventually outlawed in 1905 (Brook et al., 2008). Whether the pirates faced decapitation or death by slicing, the executions were usually expedited and carried out right after the conclusion of the trials.

Although piracy was considered a particularly egregious crime and dealt with as such, like the trials during the Golden Age of piracy, Chinese imperial courts acknowledged that not all pirates are alike and made every attempt to differentiate the "hardcore" pirates from those who were forced into piracy (Antony, 1992). Those pirates deemed to be hardcore, cutthroat criminals who knowingly and willingly participated in piracy faced decapitation or death by slicing. However, pirate accomplices who were initially kidnapped by the pirates and forced to serve as pirates were classified in one of two groups—those that directly help commission a crime faced banishment and slavery and those that performed menial tasks such as cooking and cleaning were forced to endure three years of penal servitude and several lashes with a whip. Women who found themselves facing the imperial judicial system as pirate accomplices received more space to negotiate whether they would be punished or released than their male counterparts. Recognizing that women were often the victims of kidnapping and rape at the hands of pirates, they could be spared death, banishment, servitude, and lashings if they were able to prove that they refused to obey the pirates and spent their time as prisoners held below deck. Proving this required the others on trial to serve as witnesses and substantiate their claims of innocence.

Somalia and Gulf of Guinea

The policing and prosecution of pirates operating off the coast of Somalia in contemporary times demonstrates both a reemergence of tactics from the past and the introduction of new approaches in response to the contextually specific challenges. Western ideas of policing are not necessarily the same as those employed in the Global South, particularly in Somalia. Rather, policing in Somalia "reflects a blend of political and functional imperatives, past practices, current contingencies, personal experience, societal and international expectations" that are carried out in a highly insecure, legally pluralistic, and politically unstable context (Hills, 2014: 779). The visibility of militarized police in Somalia conveys a level of preparedness that helps garner public support for police empowerment that translates into the public's willingness to listen to and abide by the police's orders and tolerate the use of police force (Muibu, 2021). Thus, Somalia's "good enough" policing plays out on a field of "good enough" governance to achieve the minimum goal of stabilization and development (Hills, 2020).

Years of civil war and unrest and failed international interventions also ushered in a complex rule of law landscape that includes four different justice systems operating throughout Somalia. Le Sage (2005) identifies the four systems as follows: (1) a formal judiciary structure in regional and central governments; (2) traditional, clan-based system known as *xeer*; (3)

shari'a courts; and (4) civil society and private sector initiatives and ad hoc militia-established mechanisms. These systems coexist and can even overlap in the same jurisdictions. International efforts to combat piracy added what should be considered a fifth justice system to the landscape—foreign trials.

During the peak of piracy off the coast of Somalia, the United Nations deemed the Somali justice system incapable of providing pirate suspects with a fair, efficient, and humane trial and imprisonment and established a regional piracy prosecution model. Within this model, the piracy suspects, all of whom held Somali citizenship, were transferred by arresting foreign navies to Kenya, the Seychelles, or Mauritius to stand trial and serve their sentences until Somalia's courts and prisons were brought up to international human rights standards. Accordingly, these foreign courts, although facilitated by the UNODC, operated much like the Admiralty Court of the Golden Age of Piracy that enabled the metropole to extend its reach and power to prosecute pirates in the colonies. Many Somali elders were vocally opposed to this approach, claiming that the punishment of Somali pirates should be done at the clan level in accordance with *xeer*. Each clan would punish the pirates of their own clan—an approach they argued would increase clan responsibility and accountability for stopping piracy. International counter-piracy groups such as the Contact Group for Piracy off the Coast of Somalia consisting of representatives from foreign governments, militaries, international agencies, and the private sectors doubted the feasibility of such an approach, arguing that instead of prosecuting and punishing the pirate, they would protect those from their own clan. Similarly, it would require all clans to buy into the approach or risk creating a piecemeal approach to justice.

None of the more than three hundred pirates arrested and prosecuted for piracy in the East Africa region were women. Rather, Somali officials quickly dismissed discussions regarding policing and punishing women pirates during counter-piracy strategy meetings, as they did not want Somali women being viewed as criminals and morally comprised, nor did they want to appear "soft" on piracy. The assumption was it was more likely that the international community would continue to invest millions in military and criminal justice programming to combat piracy in Somalia if the pirates were presented as rogue, dangerous men, than if they were known to be a bunch of women selling food and guns by the seashore (Gilmer, 2014, 2019). Despite the existence of United Nations gender equality goals and pleas from local Somali women's groups to prioritize programing that can help warn their women of the dangers of piracy and support the women who are victims of pirates, women were not considered a target group for piracy recruitment or potential offenders. Internationally funded antipiracy advocacy campaigns prohibited Somali government officials from alluding to the possibility of women pirates and, instead, were only able to portray women as potential victims of piracy (Gilmer, 2016). Somali officials were resolute that any woman involved in piracy would be dealt with by their family and clan however they saw fit rather than be paraded in front of the international presses captivated by the regional piracy prosecution trials.

The preference for dealing with piracy-involved women in private is reflective of broader social patterns in Somalia. Somali families predominantly deal with female delinquency within the family and/or clan to maintain the respectability of both the woman and her relatives. Similarly, women are prohibited from participating in the traditional *xeer* justice system, rendering their interests and representation almost entirely invisible in rule of law (Abdulkadir & Abdulkadir, 2012). Most Somali prisons do not have space for a women's wing, which also contributes to the low number of women incarcerated in Somalia. The charges that land women in prison are often related to (suspected) association with the extremist group al-Shabaab (Hassan, 2021) or professional malpractice as is the case of one female inmate who had been sentenced to life in prison because she delivered a stillborn baby (Gilmer, 2014). Other women were incarcerated for violating social expectations by alleging they were raped (Tekle, 2013) or being accused of drinking alcohol (CBC News, 2019).

Al-Shabaab supersedes local laws and oversees the policing and punishing of women (and men) in areas where they control territory, particularly South-Central Somalia. The group's enforcement of Sharia law tends to be even more strict on women's movements, behavior, and dress than Somali customary law. For example, in 2011, al-Shabaab banned women and men walking together in public if they are not married, as well as handshakes between men and women. These activities were punishable by public flogging (BBC, 2011). In the town of Baidoa, women who did not wear the proper Islamic dress faced jail time (Brumfield & Adow, 2009). A 20-year-old woman accused of adultery was stoned to death in the town of Wajid while her boyfriend was given one hundred lashes (BBC, 2009).

Despite Somalia's various justice systems that include families, clans, the government, and al-Shabaab to policing and correcting women's behaviors, there is yet to be a publicly known case of a woman arrested and punished for piracy-related activities. To date, only men have been arrested and sentenced for piracy in Somalia or within the broader regional piracy prosecution model. As of November 2012, the peak of piracy off the coast of Somalia, UNODC reported that at least 1,186 Somali *men* were facing prosecution for piracy in over 21 countries around the world (UNODC, 2012). This is also the case with piracy off the GoG in West Africa. Much of the GoG piracy is the product of the conflict surrounding the regional oil industry. Like Somalia, the rule of law, penal systems, and amnesty programs in GoG states have failed to stop piracy. Even a presidential amnesty program offered by the Nigerian president in 2009 failed to quell piracy. As piracy groups continue to ravage the high seas and riverways of the GoG, women are reportedly playing a large role in the logistical aspects of the operations. Women are embedded within the broad social framework that facilitates providing food and other services to the pirates as well as managing the day-to-day operations of the camps where piracy hostages are held (in conversation with a program officer of the UNODC Global Maritime Crime Programme Atlantic Ocean division, March 2021). Despite their known, critical involvement in pirate groups, women's active participation in piracy is not mentioned in a recent UNODC titled *Pirates of the Niger Delta* (Jacobsen, 2020).

Invisible Pirates, Invisible Punishments?

Piracy trials during the Golden Age of piracy were fast spectacles compared to the slow, closed-door hearings of contemporary Somali piracy. Most trials throughout history demonstrate that odds were against the accused and officials aggressively pursue guilty verdicts (Rediker, 2004; Simon, 2016; Gibbs, 2019). However, perhaps these general statements need to be revised to reflect gendered realities more accurately. While this may be true of *men* pirates, the same cannot be said for the arrest and punishment of *women* pirates. Records of historical piracy provide evidence of women being tried and acquitted for aiding pirates in 1500's England. Women pirates in the 1700's Americas were not as fortunate, as some were hung alongside the convicted men. Women of eighteenth- and nineteenth-century piracy in China skillfully negotiated amnesty and drew on their political connections to evade arrest and punishment. The women of contemporary piracy off the coast of Somalia and in the GoG have yet to be targeted with punitive measures.

Women pirates do not seem to have been afforded the same label, *hostis humani generis* (enemies of mankind) as their male counterparts. Whereas the action of male pirates was considered so far beyond the pale of civilized society, and such a threat to statehood, that governments were encouraged to "destroy him by foul means", the action of piracy-involved women has continually been delegitimized, trivialized, and even actively ignored (Baer, 1982: 7). Women pirates seem to be aware of this gendered exceptionalism and "do femininity" if/when they are faced with the criminal justice system. As evidenced in the opening anecdote of Anne Bonny and Mary Read, they understood the power of their sexed bodies when they shouted out their pregnancies hoping for a stay of execution. The male gaze of justice designated their pregnant bodies as "out of order for punishing" (Langan et al., 2019), which was later resumed when they were no longer with child. Similarly, the piracy-involved women of Somalia capitalize on the averted gaze of men and their government to carry out their tasks, increase their profits and those of pirates, and evade punishment.

Contemporary women pirates of Somalia and the GoG states possess and employ a form of "street capital" much like women drug dealers (Grundetjern & Sandberg, 2012) that involves technical knowledge of local markets and an awareness and willingness to exploit gendered expectations and cultural norms at multiple scales—the local, state, and international to successfully carry out their missions. In doing so, they appear to coopt the international system that treats women as representatives of clan honor (Abdi, 2007) and as depositories of culture (Yuval-Davis, 1997) to evade the pitfalls and stereotypes that plague women who participate in political violence. For example, women involved in terrorist activities are seen as more deranged and threatening than their male counterparts and are abused and humiliated by law enforcement (Bloom, 2011). This is not to suggest that women pirates should receive similar or harsher punishments than other women involved in transnational crime and political violence. Rather, it shows a historical continuity of women pirates being treated as the *exception* by criminal justice systems. Women's central role in contemporary piracy suggests the need for a closer examination about why women pirates continue to be rendered invisible. Who benefits from their invisibility and how? The remainder of the chapter examines the women that are allowed to be seen—those serving as counter-piracy subjects.

Counter-Piracy Subjects

As the prop plane circled the Garowe airport, a small group of dignitaries, media reporters, and security agents stood huddled in front of the main building. On this Sunday, everyone had come to oversee the arrival of two of Somalia's recently captured pirates. These pirates had elicited an emergency repatriation from a prison in the Seychelles to Puntland, Somalia. The anticipation and trepidation could be felt as the wheels of the plane finally touched down on the dusty, desolate runway. As the plane door bounced open, the steps unfolded onto the ground and two faces emerged from the opening. Murmurs of excitement arose throughout the crowd that had now pushed its way close enough to the plane to gaze upon the pirates. Everyone watched as one of the pirates, a 13-year-old boy, descended the steps and ran into the arms of an elder woman and a younger woman by her side—his mother and sister.

The three were only allowed a brief embrace before they were rushed off to the main room of the airport where the group of reporters were now anxiously waiting to get their photo opportunity. The two boys and two women were ushered to take their place at a long table adorned with outdated microphones. The boys sat with their hands in their lap evading eye contact with everyone in the room as the regional counter-piracy director waxed poetic about his office's joint efforts with the international community to bring an end to piracy in Puntland. The media left the boys alone and instead turned their energies toward the mother and began peppering her with questions. She recoiled. Her nervousness was palpable as she struggled to project her quiet, trembling voice. The sister kept her hand on her mother's arm as she spoke as if to both steady her and reassure her that the interview would soon be over. As the mother finished what felt like a pre-scripted explanation of how her son was recruited into piracy, the counter-piracy director grabbed the microphone nearest him, looked directly at the one video camera in the room, and insisted upon the importance of mothers telling their sons of the dangers of piracy and keeping them away from pirates.

The woman perched in front of journalists at the airport that day became the first of many "security mothers" and "sisters" (Grewel, 2006; Fixmer-Oraiz, 2019) called upon to advocate for protecting sons (and husbands) against the scourge of Somali piracy. Contrary to the many media images depicting Somali pirates as war-hardened, grown men draped in belts of ammunition with an AK-47 in their hand, when in reality, many of them were young, impoverished, frightened men who gravely regretted the situation they found themselves in. Thus, the mothers and wives of Somali men came to serve a vital counter-piracy role—albeit predominantly from the privacy of their own homes—as they were enlisted by the government and international agencies to help combat the recruitment of young men into piracy.

Instrumentalizing Motherhood for Safety and Security

The gendered way in which women have been employed to aid in the suppression of piracy, and the instrumentalization of motherhood, is not unique to counter-piracy programming efforts. Feminist political scholars continue to critique the use of women as symbols of the nation by leaders and intellectuals (Keohane, 1989; Bracewell, 1996; Ranchood-Nillson & Tetreault, 2000; Enloe, 2014; Milic, 2018). The use of women as symbols diminishes their agency and invalidates any meaningful participation; they solely become *vehicles* for transmitting the whole nation's values from one generation to the next as the "*bearers* of the community's future generations—or, crudely, nationalist wombs" (Enloe, 2014: 108; emphasis in original). Feminist political geographers have also theorized the ways in which the nuclear family has become politicized and recast as a model for political relations (Cowen & Gilbert, 2008). Within this model, women's role as mother and wife can become securitized, and even militarized, as a tool for responding to national challenges such as terrorism and piracy (Cowen & Gilbert, 2016).

The instrumentalization of motherhood for the purposes of safety and security gained public and scholarly attention in the aftermath of the 9/11 terrorist attacks in the United States. Grewal (2006) identified the emergence of "security moms" as a key voting bloc in the 2004 American elections. Based on an editorial column written by a self-described married, gun owning, voting, mother of two children, "security moms" were female citizen-subjects who prioritized the safety of their home and the survival of their homeland. These women were encouraged to monitor those around them, pay attention to suspicious activity, become familiar with the Federal Bureau of Investigation most wanted list, and educate their children on security threats (e.g., Muslims and illegal aliens). To Grewal (2006), the emergence of security moms signals a collaboration between public and private patriarchies that promotes the view of children as innocent victims and encourages nationalism, increased surveillance, and the militarization of the home/mothers. Mothering as a set of beliefs and practices, thus, becomes co-opted as a security function with underlying expectations that the proper gendered female subject of the home is inherently antiterrorism and anticrime. Additional research on security moms reveals that both mothers *and* childless women supported the militarization of the home in post-9/11 America (Elder & Greene, 2007).

The security mom subject is no longer contained within the imaginaries and politics of American society. Mothering as a security function has since been reimagined and reinforced globally. The contemporary security mom has been internationalized in that she is developed and supported through international mechanisms such as United States Agency for International Development and United Nations agencies. The internationalized security mom of the Global South is encouraged to *de*militarize her home rather than militarize it. She is on a peace mission to identify and root out potential violent extremists and terrorists whose actions may threaten not only the safety and security of her country but that of the United States and other Western nations. Accordingly, she is responsibilized with encouraging a peaceful, pacifist agenda and making sure her children and husband do not become the threating "others" that may harm Western populations and interests.

Countering Violent Extremism (CVE) programs worldwide have operationalized the social feminist' argument that mothering fosters empathy, caring, and pacifist attitudes under the guise of promoting gender equality and women's empowerment (Elshtain, 1981; Ruddick, 1989, 2018; Keller, 2010; Mayseless, 2015; Dunkerley, 2017). United Nations Development Programs (UNDP) aimed at preventing violent extremism in Pakistan identify educated and financially independent women as central to being able to recognize the risk of radicalization in children and opposing fundamentalist narratives (UNDP, 2016). Similar CVE programs in Bangladesh employ a dual approach of improving women's economic position while also training them to identify and intervene intolerance that leads to radicalization in their rural communities and local schools (Henty & Eggleston, 2018). The Mothers School in Tajikistan capitalizes on existing local mother's group to provide resources and support such as teaching communication skills, parenting skills, and strategies for becoming more involved with their children's education as a means for identifying warning signs of radicalization.

The clear emphasis these programs place on the link between "good mothering" and the ability to prevent radicalization and violent extremism perpetuates the view that women are the foundation of household, community, and national safety and security (Gasztold, 2020). Identifying and instrumentalizing mothers as a "first line of defense" unduly politicizes the mother–child relationship and pushes the responsibility for national security down onto the private sphere and a woman's ability to "properly" nurture her children (and husband). Thus, because of their position within the family, mothers become key informants and actors in monitoring the "social and psychological landscape" of those in the home to intervene at an early stage to prevent involvement in violent extremism (Schlaffer & Kropiunigg, 2016; Henty & Eggleston, 2018).

Blurring the Boundary between Piracy and Terrorism

The merging of terrorism and piracy in discourse, policy, and practice contributed to a situation ripe for the employment of Somali women, as mothers, in the fight to suppress piracy. First, though it is important to understand that the merging of piracy and terrorism is not a contemporary phenomenon. Historically, piracy was a state-commissioned crime called "privateering" where "pirates/privateers" operated under state-issued licenses and were granted permission by the issued state to attack and plunder enemy ships. Eventually, privateering was outlawed and deemed illegal after states were able to successfully build up their economies and accompanying navies. Thus, piracy's roots began with a complex entanglement of economic and political interests. That entanglement has reemerged in the post-9/11 era as government leaders and military officials have attempted to tie pirates to terrorist groups operating in the regions (Thorup, 2009). For example, piracy in Southeast Asia has been tied to the Moro Islamic Liberation Front and

Abu Sayyaf Group (Raymond, 2006), piracy in the GoG has been attributed to the disorder created by the Movement for the Emancipation of the Niger Delta (Nwalozie, 2020), and piracy off the coast of Somalia to al-Shabaab (Levy & Yusuf, 2021).

Samatar et al. (2010) further complicate homogenizing understandings of the history and root causes of piracy off the coast of Somalia. They provide a convincing argument that piracy off the coast of Somalia began as a politically motivated crime carried out between domestic-based groups vying for control over the federal government. It eventually morphed into several other types of piracy including resource and defensive piracy and eventually became the ransom piracy that dominated the waters during Somalis "piracy crises" of 2009–2012. Journalists and scholars who interviewed piracy prisoners in the Seychelles confirmed that ransom pirates are motivated by the prospect of personal financial gain and cultural expectations linked to economic realities in Somalia (UNODC, 2015; Gilmer, 2017). A similar study conducted with piracy prisoners in Kenya suggests that piracy off the coast is strictly an economic crime (Hamilton, 2011).

Despite ransom piracy's well-documented economic motivations, the criminal enterprise has been reconceived as economic terrorism (Singh Parmar, 2012) and strategically politicized by foreign governments and international agencies. Singh and Bedi (2016) contend that foreign governments have preyed upon the public's vulnerabilities by exaggerating the link between pirates and al-Shabaab and al-Qaeda and portraying them as "maritime terrorists". Similarly, Engels' (2007) critical analysis of the wording of the National Security Strategy of the United States astutely highlights how descriptions of piracy and terrorism are conflated and interchanged so that the piracy theft is reimagined as floating bombs that necessitate military response. Doing so allows them to facilitate extraordinary legal measures and militarize the waters off the coast of Somalia and broader Indian Ocean region (Gilmer, 2014; Sing & Bedi, 2016). Young and Valencia (2003) argue that a similar process is taking place in Southeast Asian waters where a push for foreign military intervention and the adoption of international maritime laws have raised issues of sovereignty.

Somali Women as Counter-Piracy Moms and Performers

Previous chapters highlighted Somali official's opposition to considering and treating Somali women as potential pirates. Although the women of Somalia were not considered part of the piracy *problem*, they were considered part of the *solution*—albeit in very niche ways. They became both the subjects and the objects of counter-piracy programming in their perceived roles as "good mothers" and "good stewards of Somali culture". The remaining sections will explore how Somali women were talked about and what they were enlisted to help with as part of the counter-piracy advocacy campaign.

The use of print/media by states to suppress piracy is not unique to contemporary piracy off the coast of Somalia. Burgess' (2009) research on the Henry Every trials of 1696 demonstrates a concerted effort by the English government to manipulate information through the media in order to turn the public against pirates. In particular, the government made the rape of women captives aboard the raided *Ganj-i-Sawai* central to the narrative of seeking justice for victims of piracy in hopes of restoring diplomatic relations with India and turning the public against pirates (Burgess, 2009). The deliberate use of print culture to win over the public underscores how critical the masses are to suppressing piracy. Using messaging to make the need to protect women from piracy synonymous with the need to protect the state piracy, then, has been around for at least three hundred years.

Marriage and family are central to Somali culture. Within a Somali marriage a man is expected to provide for his family and the woman is expected to care for him and the children. Accordingly, marriage and women's work in the home are seen as crucial to social reproduction and the maintenance of social, economic, and political organization in Somali society (Ahmed, 2004). In 2012, during the peak of piracy off the coast of Somalia, foreign governments funded a UNODC-facilitated antipiracy campaign designed to reflect and amplify the strongly defined rules and customs that make up Somali culture and traditions to turn the Somali public against pirates (Bueger, 2012; Gilmer, 2016). Of the four messaging narratives, two were specifically aimed at highlighting the impact of piracy on Somali society and families: (1) piracy is ruining Somali culture and traditions, and (2) pirates who go to sea are at risk of death and/or imprisonment.

The first message, piracy is ruining Somali culture and traditions, similar to the Every trial, sought to make the condition of Somali women synonymous with the condition of Somali culture and traditions. Somali officials took the opportunity to blame the growth of particular social ills such as increased prostitution and women drinking and disobeying their families solely on pirates. While there is certainly correlation between the two, the government was able to use the campaign messages to scapegoat pirates for these vices rather than acknowledging the possibilities of a society changing alongside broader global socioeconomic and cultural change. In practice, however, the "keep-your-daughters-sisters-and-mothers-safe-from-pirates" directive fell to the women to operationalize within their home.

Rather than addressing the possibilities that women may freely *choose* to prostitute themselves to or marry pirates, the messaging campaign targeted men to control their urges and portrayed sex outside of marriage as dirty, un-Islamic, and therefore not in line with Somali values. Somali religious leaders were tasked with addressing prostitution and infidelity in mosques and civil groups by encouraging men to protect themselves from and refrain from prostitution. Focusing on the "moral shortcomings" of the Somali men left the image of Somali women unscathed. Feminist theorist Narayan (1998) labels this practice as "selective labeling", whereby those with social power conveniently designate certain changes in values and practices as consonant with "cultural preservation" while designating other changes as "cultural loss" or "cultural betrayal" (8–9). Women, however, were instructed to privately discourage Somali girls from pursing relationships with pirates—something that was left out of the public face of the campaign.

In addition to dissuading their daughters from seeking out relations with pirates, Somali women were encouraged to take on the role of counter-piracy moms by preventing their sons and husbands from joining pirate groups. The other campaign message, pirates who go to sea are at risk of death and/or imprisonment, was designed to scare Somali women into protecting their sons and husbands. Somali officials and UNODC program officers contemplated broadcasting radio announcements of men who died at sea or were imprisoned in nearby states to elicit a response from Somali women—as if they had not already been desensitized from losing their sons and husbands to years of civil war and violent extremism. Those women who did lose their men to piracy were encouraged to speak out about their grief and hardships as evidenced in the chapter's opening anecdote. These women served a dual purpose—they became the face of Somalia's counter-piracy mothers that would encourage other mothers to do their part, and they sent a message to current and potential pirates that your crimes are hurting your mothers and wives.

The messaging campaign conflated preserving Somali culture with the creation and maintenance of traditional, pirate-free households and Somali communities. The messaging discourse removed the responsibility for security from the government and placed it in the home, the space where Somali women earn their respect and value in society. Like the women recruited into CVE programs, counter-piracy moms were deemed most likely to be able to monitor suspicious behavior and intervene if/when their sons and husbands appeared to be considering joining piracy groups. For those whose men had already been recruited into piracy, they were encouraged to call them back. Lastly, they were to ensure the preservation of their daughters' dignity, and by extension the dignity of Somalia, by educating them on the realities of being in a relationship with pirates and the likelihood of abuse, drug use, and familial and social exile. These fraudulent and dangerous relationships were identified as the top concern by Somali women who pushed, unsuccessfully, to have the messaging included in the antipiracy campaign (Gilmer, 2014). Accordingly, the women were encouraged to take it upon themselves to ensure the truth about pirates was being communicated to young Somali women.

Women were allowed and encouraged to occupy the public roles of counter-piracy performers. Somali officials and UNODC staff envisioned a counter-piracy caravan that would travel throughout Somalia putting on plays, leading songs, and delivering speeches about the dangers of piracy. Although security risks in the rural areas put a stop to the caravan before it started, smaller advocacy performances were eventually held in some of Somalia's urban areas through various United Nations agencies funded by the Contact Group on Piracy off the coast of Somalia. These events featured women performing plays, poetry, song, and dance that delivered antipiracy messages (Contact, 2010). Thus, women were called upon to simultaneously embody Somali culture and counter-piracy efforts through these public performances such as the ribbon-cutting ceremony for the construction of the UNODC CPP-funded Garowe Prison project with foreign dignitaries and the executive director of UNODC, Yury Fedotov, in attendance (UNODC, 2012).

As piracy attacks off the coast of Somalia dwindled after 2013, the naval operations remained but the funding for onshore advocacy programs eventually dried up. Conversations about the importance of Somali women and mothers in the fight against piracy, however, continued among policymakers and counter-piracy practitioners. As recently as 2017, representatives from the NGO, Oceans Beyond Piracy, presented a working paper at the Women Peace & Security conference held at the U.S. Naval War College in Newport, Rhode Island. They argue that women and especially mothers are an essential part of Somali society that can act as a channel of dialogue between families and clans (Lawellin & Monahan, 2017). For example:

> Mothers have more direct contact with children under the age of 15 giving them a unique role in combating Somali piracy. This points out the unique role that mothers can have in deterring Somali piracy as they can have a persuasive effect in deterring youth from becoming pirates. Engaging with mothers can be an effective method in building deterrence strategies.
>
> (51)

They further cite the "success" mothers have had in CVE programming in Somalia as evidence that counter-piracy programming can also capitalize on a mother's unique role and "empower" them to prevent their children and husbands from going down the path of piracy. They acknowledge that previous UNODC antipiracy advocacy campaigns only targeted messaging toward Somalia men and proposed a women-focused messaging campaign that includes messages such as "Don't Marry a Pirate".

Although it is encouraging to see recognize the potentially important impacts women can have on counter-piracy efforts, it is imperative that women are approached and involved as active participants rather than subjects. Programs that rely heavily on gendered logics can be detrimental to women by essentializing them and preventing them from realizing their agency and rights. For example, Brown's (2013) groundbreaking comparative study of women's participation in counterterror measures in the United Kingdom, Indonesia, and Saudi Arabia found that Muslim women often become counter-piracy *subjects* rather than *agents* through a process that solely understands them through their relations with male relatives, assumes they are guided by maternal instincts that promote peace and oppose violence, and essentializes their participation as evidence that they are practicing "good Islam". Or, to simplify—good mothers do not produce radicals.

There is a danger in assuming what works for CVE programming will work for counter-piracy programming. More effort needs to be made to ensure that these programs do not implicitly assume that women support the crime prevention initiatives or are peaceful in their views. There also needs to be pushback on constructions of women as inherently relatives, wives, and mothers of active and potential criminals and extremists. Women's empowerment should not be, as Brown so brilliantly describes, "simply a vehicle to access 'vulnerable men'" (45). Counter-piracy work should not be made "mother work", and programming must avoid suggesting that good mothers do not produce pirates. Gayatri Spivak once asserted that war is rationalized as white men seeking to save brown women from brown men (1999, see also Nast, 2000). Ironically, CVE and counter-piracy programming appear to be presenting a twist on this phenomenon where white (Western) men are seeking brown women to save them from brown men (from the Global South).

It would be remiss to not conclude this chapter by acknowledging the many ways women have been and are currently involved in combating piracy. Although their numbers are far fewer than men, women play instrumental roles as sailors aboard foreign vessels apprehending pirates, as policymakers working within state offices, or in various United Nations agencies helping to craft and implement counter-piracy policies and programs. These women work tirelessly to assist victims of piracy in Somalia, to understand factors that motivate pirates in Southeast Asia, to suppress piracy in the GoG, and to prosecute and imprison those found guilty of piracy around the world. To all the women who are actively working on and researching about counter-piracy issues … I see you, and I thank you.

References

Abdi, C. M. (2007). Convergence of civil war and the religious right: Reimagining Somali women. *Signs: Journal of Women in Culture and Society, 33*(1), 183–207.

Abdulkadir, F., & Abdulkadir, R. (2012). Gender & transitional justice: An empirical study on Somalia. *Role of Women in Promoting Peace and Development, 161*, 161–178.

Ahmed, S. M. (2004). Chapter 2: Traditions of marriage and the household. In J. el Bushra & J. Gardner (Eds.), *Somalia –The untold story: The War through the Eyes of Somali Women* (pp. 51–68). Sterling, VA: Pluto.

Antony, R. J. (1992). The suppression of pirates in South China in the mid-Qing period. *American Journal of Chinese Studies, 1*(1), 95–121.

Antony, R. J. (2012). Bloodthirsty pirates? Violence and terror on the South China sea in early modern Times1. *Journal of Early Modern History, 16*(6), 481–501.

Appleby, J. C. (2013). *Women and English piracy, 1540–1720: Partners and victims of crime*. Suffolk: Boydell & Brewer Ltd.

Appleby, J. C. (2016). Pirates and communities: Scenes from Elizabethan England and Wales. In J. Appleby & P. Dalton (Eds.), *Outlaws in medieval and early modern England* (pp. 149–172). Oxfordshire: Routledge.

Atkins, S., & Brenda Hale, B. (2018). *Women and the law*. London: University of London, p. 284.

Baer, J. H. (1982). "The complicated plot of piracy": Aspects of English criminal law and the image of the pirate in defoe. *Eighteenth Century, 23*(1), 3–26.

Balfour, G., & Comack, E. (Eds.). (2021). *Criminalizing women: Gender and (in) justice in neoliberal times*. Winnipeg: Fernwood Publishing.

(BBC) British Broadcasting Corporation News. (2009, November 18). Somali woman stoned for adultery. Retrieved from http://news.bbc.co.uk/2/hi/africa/8366197.stm

(BBC) British Broadcasting Corporation News. (2011, January, 7). Somalia's al-Shabab bans mixed-sex handshakes. Retrieved from https://www.bbc.com/news/world-africa-12138627

Benton, L. (2011). Toward a new legal history of piracy: Maritime legalities and the myth of universal jurisdiction. *International Journal of Maritime History, 23*(1), 225–240.

Bloom, M. (2011). Bombshells: Women and terror. *Gender Issues, 28*(1), 1–21.

Bracewell, W. (1996, January). Women, motherhood, and contemporary Serbian nationalism. *Women's studies international forum* (Vol. 19, No. 1–2, pp. 25–33).

Brook, T., Bourgon, J., & Blue, G. (2008). *Death by a thousand cuts*. Cambridge, MA: Harvard University Press.

Brown, K. E. (2013). Gender and counter-radicalization: Women and emerging counter-terror measures. In K. E. Brown (Ed.), *Gender, national security, and counter-terrorism* (pp. 49–72). Oxfordshire: Routledge.

Brumfield, B., & Adow, M. A. (2009, April 20). Women in Somali city must cover up or go to jail. Retrieved from https://www.cnn.com/2009/WORLD/africa/04/20/somalia.islamic.law/

Bueger, C. (2012). Drops in the bucket? A review of onshore responses to Somali piracy. *WMU Journal of Maritime Affairs*, *11*(1), 15–31.

Burgess, D. R. (2009). Piracy in the public sphere: The Henry Every trials and the battle for meaning in seventeenth-century print culture. *Journal of British Studies*, *48*(4), 887–913.

Carrington, K., McIntosh, A., & Scott, J. (2010). Globalization, frontier masculinities and violence: Booze, blokes and brawls. *British Journal of Criminology*, *50*(3), 393–413.

(CBC) Canadian Broadcasting Company. (2019, May 12). 2 Canadian women freed from Somaliland prison say they endured extreme abuse. Retrieved from https://www.cbc.ca/news/canada/toronto/two-canadian-women-jailed-somaliland-alcohol-return-canada-1.5132887

Chesney-Lind, M. (2002). Criminalizing victimization: The unintended consequences of pro-arrest policies for girls and women. *Criminology and Public Policy*, *2*(1), 81–90.

Chesney-Lind, M. (2017). Policing women's bodies: Law, crime, sexuality, and reproduction. *Women and Criminal Justice*, *27*(1), 1–3.

Chesney-Lind, M., & Morash, M. (2013). Transformative feminist criminology: A critical re-thinking of a discipline. *Critical Criminology*, *21*(3), 287–304.

Collins, V. E., & Dunn, M. (2018). The invisible/visible claims to justice: Sexual violence and the university camp (us). *Contemporary Justice Review*, *21*(4), 371–395.

Contact group on piracy off the coast of Somalia. (2010). Trust fund to support the initiatives of states countering piracy off the coast of Somalia. Retrieved from https://www.unodc.org/documents/easternafrica/piracy/Annual_Report_2010_Piracy_TF_eng_eBook.pdf

Cook, K. J. (2016). Has criminology awakened from its "androcentric slumber"? *Feminist Criminology*, *11*(4), 334–353.

Cordingly, D. (2006). *Under the black flag: The romance and the reality of life among the pirates*. New York: Random House Trade Paperbacks.

Cowen, D., & Gilbert, E. (2008). Citizenship in the "Homeland". In D. Cowen & E. Gilbert (Eds.), *War, citizenship, territory* (pp. 261–279). Oxfordshire: Routledge.

Cowen, D., & Gilbert, E. (2016). Fear and the familial in the US war on terror. In R. Pain & S. J. Smith (Eds.), *Fear: Critical geopolitics and everyday life* (pp. 67–76). Oxfordshire: Routledge.

Creighton, M. S., & Norling, L. (Eds.). (1996). *Iron men, wooden women: Gender and seafaring in the Atlantic World, 1700–1920*. Baltimore: JHU Press.

Daly, K., & Chesney-Lind, M. (1988). Feminism and criminology. *Justice Quarterly*, *5*(4), 497–538.

DeKeseredy, W. (2000). *Women, crime and the Canadian criminal justice system*.Brisbane: Queensland University of Technology.

Dunkerley, S. (2017). Mothers matter: A feminist perspective on child welfare-involved women. *Journal of Family Social Work*, *20*(3), 251–265.

Elder, L., & Greene, S. (2007). The myth of "security moms" and "NASCAR dads": Parenthood, political stereotypes, and the 2004 election. *Social Science Quarterly*, *88*(1), 1–19.

Elshtain, J. B. (1981). *Public man, private woman*. Princeton, NJ: Princeton University Press.

Engels, J. (2007). Floating bombs encircling our shores: Post-9/11 rhetorics of piracy and terrorism. *Cultural Studies? Critical Methodologies*, *7*(3), 326–349.

Enloe, C. (2014). *Bananas, beaches and bases: Making feminist sense of international politics*. Oakland: University of California Press.

Fixmer-Oraiz, N. (2019). Homeland maternity: US security culture and the new reproductive regime. Champaign: University of Illinois Press.

Gaarder, E., & Belknap, J. (2002). Tenuous borders: Girls transferred to adult court. *Criminology*, *40*(3), 481–517.

Gasztold, A. (2020). Feminist perspectives on Terrorism. New York: Springer International Publishing.

Gibbs, J. (2019). The brevity and severity of 'golden age' piracy trials. *International Journal of Maritime History*, *31*(4), 729–786.

Gilmer, B. (2014). *Political geographies of piracy: Constructing threats and containing bodies in Somalia*. Berlin: Springer.

Gilmer, B. (2016). Awareness campaigns as a response to piracy off the coast of Somalia. *Journal of Development Studies*, *52*(6), 765–779.

Gilmer, B. (2017). Hedonists and husbands: Piracy narratives, gender demands, and local political economic realities in Somalia. *Third World Quarterly*, *38*(6), 1366–1380.

Gilmer, B. (2019). Invisible pirates: Women and the gendered roles of Somali piracy. *Feminist Criminology*, *14*(3), 371–388.

Grewal, I. (2006). "Security moms" in the early twentieth-century United States: The gender of security in neoliberalism. *Women's Studies Quarterly*, *34*(1/2), 25–39.

Hamilton, K. (2011). The piracy and terrorism nexus: Real or imagined? *Journal of the Australian Institute of Professional Intelligence Officers*, *19*(2), 25–37.

Grundetjern, H. (2015). Women's gender performances and cultural heterogeneity in the illegal drug economy. *Criminology*, *53*(2), 253–219.

Grundetjern, H., & Sandberg, S. (2012). Dealing with a gendered economy: Female drug dealers and street capital. European Journal of Criminology, 9(6), 621–635.

Hassan, A. (2021, March 5). Al Shabaab militants storm Somali jail, seven soldiers killed. *Reuters*. Retrieved from https://www.reuters.com/article/us-somalia-violence-idUSKBN2AX188

Henty, P., & Eggleston, B. (2018). Mothers, mercenaries and mediators: Women providing answers to the questions we forgot to ask. *Security Challenges*, *14*(2), 106–123.

Hills, A. (2014). What is policeness? On being police in Somalia. *British Journal of Criminology*, *54*(5), 765–783.

Hernandez, C. M. (2009). Forging an Iron Woman: On the effects of piracy on gender in the 18th century Caribbean. Vanderbilt Undergraduate Research Journal, 5.

Hills, A. (2020). The dynamics of prototypical police forces: Lessons from two Somali cities. *International Affairs*, *96*(6), 1527–1546.

Irwin, K., Pasko, L., & Davidson, J. T. (2018). Girls and women in conflict with the law. In W. DeKeseredy & M. Dragiewicz (Eds.), *Routledge handbook of critical criminology* (pp. 358–368). Oxfordshire: Routledge.

Jacobsen, K. L. (2020). Pirates of the Niger Delta. United Nations Office on Drugs and Crime Global Maritime Crime Programme. Retrieved from https://www.unodc.org/res/piracy/index_html/UNODC_GMCP_Pirates_of_the_Niger_Delta_between_brown_and_blue_waters.pdf

Keller, J. (2010). Rethinking Ruddick and the ethnocentrism critique of maternal thinking. *Hypatia*, *25*(4), 834–851.

Keohane, R. O. (1989). International relations theory: Contributions of a feminist standpoint. *Millennium*, *18*(2), 245–253.

Kwan, C. N. (2020). In the business of piracy: Entrepreneurial women among Chinese pirates in the mid-nineteenth century. In J. Aston & C. Bishop (Eds.), *Female entrepreneurs in the long nineteenth century* (pp. 195–218). New York: Palgrave Macmillan.

Langan, D., Sanders, C. B., & Gouweloos, J. (2019). Policing women's bodies: Pregnancy, embodiment, and gender relations in Canadian police work. *Feminist Criminology*, *14*(4), 466–487.

Lawellin, B., & Monahan, M. (2017) Don't marry a pirate: Women's groups deterrence of pirate action groups in Somalia. Working Papers (August 10–11, 2017) The Next Decade: Amplifying the Women, Peace and Security Agenda. Women Peace Security. U.S. Naval War College. 2017 Conference. Retrieved from https://digital-commons.usnwc.edu/cgi/viewcontent.cgi?article=1000&context=wps

Le Sage, A. (2005). Stateless justice in Somalia: Formal and informal rule of law initiatives. Geneva: Center for Humanitarian Dialogue.

Levy, I., & Yusuf, A. (2021). How do terrorist organizations make money? Terrorist funding and innovation in the case of Al-Shabaab. *Studies in Conflict and Terrorism*, *44*(12), 1167–1189.

Lopez, V., & Pasko, L. (Eds.). (2021). *Latinas in the criminal justice system: Victims, targets, and offenders* (Vol. 18). New York: NYU Press.

MacKay, J. (2013). Pirate nations: Maritime pirates as escape societies in late imperial China. *Social Science History*, *37*(4), 551–573.

Mallicoat, S. L. (2007). Gendered justice: Attributional differences between males and females in the juvenile courts. *Feminist Criminology*, *2*(1), 4–30.

Mayseless, O. (2015). *The caring motivation: An integrated theory*. Oxford: Oxford University Press.

Milić, A. (2018). Women and nationalism in the former Yugoslavia. In N. Funk & M. Mueller (Eds.), *Gender politics and post-communism* (pp. 109–122). Oxfordshire: Routledge.

Morris, A. (1987). *Women, crime, and criminal justice*. Oxford: Blackwell.

Muibu, D. (2021). Police empowerment and police militarisation in times of protracted conflict: Examining public perceptions in southern Somalia. *South African Journal of International Affairs*, *28*(2), 233–261.

Murray, D. (1981). One woman's rise to power: Cheng I's wife and the pirates. *Historical Reflections/Réflexions Historiques*, *8*(3), 147–161.

Narayan, U. (1998). Essence of culture and a sense of history: A feminist critique of cultural essentialism. *Hypatia*, *13*(2), 86–106.

Nast, H. J. (2000). Mapping the "unconscious": Racism and the oedipal family. *Annals of the Association of American Geographers*, *90*(2), 215–255.

(The) National Archives. (2022). Support for accused. Retrieved from https://www.nationalarchives.gov.uk/education/resources/early-modern-witch-trials/support-for-accused/

National Park Service. (2020). Did you know: The US has a history of women pirates? Retrieved from https://www.nps.gov/articles/dyk-women-pirates-in-the-usa.htm

Nwalozie, C. J. (2020). Exploring contemporary sea piracy in Nigeria, the Niger Delta and the Gulf of Guinea. *Journal of Transportation Security*, *13*(3), 159–178.

Pasko, L. (2010). Setting the record "straight": Girls, sexuality, and the juvenile correctional system. *Social Justice*, *37*(1), 7–26.

Pasko, L., & Chesney-Lind, M. (2018). A critical examination of girls' violence and juvenile justice. In W. DeKeseredy & M. Dragiewicz (Eds.), *Routledge handbook of critical criminology* (pp. 348–357). Oxfordshire: Routledge.

Pasko, L., & Lopez, V. (2018). The Latina penalty: Juvenile correctional attitudes toward the Latina juvenile offender. *Journal of Ethnicity in Criminal Justice*, *16*(4), 272–291.

Penal Reform International. (2021). Global prison trends 2021. Retrieved from https://cdn.penalreform.org/wp-content/uploads/2021/05/Global-prison-trends-2021.pdf

Quinlan, C. (2017). Policing women's bodies in an illiberal society: The case of Ireland. *Women and Criminal Justice*, *27*(1), 51–72.

Ranchod-Nilsson, S., & Tétreault, M. A. (2000). *Women, states and nationalism: At home in the nation*. Oxfordshire: Routledge.

Raymond, C. Z. (2006). Maritime terrorism in Southeast Asia: A risk assessment. *Terrorism and Political Violence*, *18*(2), 239–257.

Rediker, M. (1996). Liberty beneath the Jolly Roger. In M. S. Creighton & L. Norling (Eds.), *Iron men, wooden women: Gender and seafaring in the Atlantic World, 1700–1920* (pp. 1–33). Baltimore: JHU Press.

Rediker, M. (2004). *Villains of all nations: Atlantic pirates in the golden Age*. Boston, MA: Verso.

Roychowdhury, P. (2020). *Capable women, incapable states: Negotiating violence and rights in India*. Oxford: Oxford University Press.

Ruddick, S. (1989). *Maternal thinking: Towards a politics of Peace*. Boston: Beacon Press.

Ruddick, S. (2018). Notes toward a feminist peace politics. In A. Herrmann & A. Stewart (Eds.), *Theorizing feminism* (pp. 196–214). Oxfordshire: Routledge.

Samatar, A. I., Lindberg, M., & Mahayni, B. (2010). The Dialectics of Piracy in Somalia: the rich versus the poor. *Third World Quarterly*, *31*(8), 1377–1394.

Schaffner, L. (2006). *Girls in trouble with the law*. Rutgers University Press.

Schlaffer, E., & Kropiunigg, U. (2016). A new security architecture: Mothers included! In N. C. Fink, S. Zeiger, & R. Bhulai (Eds.), *A man's world? Exploring the roles of women in countering terrorism and violent extremism* (pp. 54–75). Abu Dhabi: Global Center on Cooperative Security.

(The) Sentencing Project. (2022). Incarcerated women and girls. Retrieved from https://www.sentencingproject.org/publications/incarcerated-women-and-girls/

Simon, R. A. (2016). The problem and potential of piracy: Legal changes and emerging ideas of colonial autonomy in the early modern British Atlantic, 1670–1730. *Journal for Maritime Research*, *18*(2), 130.

Singh Parmar, S. (2012). Somali piracy: A form of economic terrorism. *Strategic Analysis*, *36*(2), 290–303.

Singh, C., & Bedi, A. S. (2016). War on piracy: The conflation of Somali piracy with terrorism in discourse, tactic, and law. *Security Dialogue*, *47*(5), 440–458.

Sjoberg, L., & Gentry, C. E. (2007). *Mothers, monsters, whores: Women's violence in global politics*. London: Zed Books.

Smart, C. (1995). *Law, crime and sexuality: Essays in feminism*. Los Angeles: Sage.

Snider, L. (2003). Constituting the punishable woman: Atavistic man incarcerates postmodern woman. *British Journal of Criminology*, *43*(2), 354–378.

Spivak, G. C. (1999). *A critique of postcolonial reason: Toward a history of the vanishing present*. Cambridge, MA: Harvard University Press.

Spohn, C., Gruhl, J., & Welch, S. (1987). The impact of the ethnicity and gender of defendants on the decision to reject or dismiss felony charges. *Criminology*, *25*(1), 175–192.

Tekle, T.-A. (2013, November 24). Rights group urges release of jailed Somali rape victim and journalist. *Sudan Tribune*. Retrieved from https://sudantribune.com/article47893/.

Thorup, M. (2009). Enemy of humanity: The anti-piracy discourse in present-day anti-terrorism. *Terrorism and Political Violence*, *21*(3), 401–411.

(UNDP) United Nations Development Programme. (2016). *Preventing violent extremism through promoting inclusive development, tolerance and respect for diversity: A development response to addressing radicalisation and violent extremism*. New York: UNDP.

(UNODC) United Nations Office on Drugs and Crime. (2012 November). Piracy is a complex challenge for East Africa's fragile economies, says UNODC executive director. Press release. Retrieved from https://www.unodc.org/unodc/en/press/releases/2012/November/piracy-is-a-complex-challenge-for-east-africas-fragile-economies-says-unodc-executive-director.html

(UNODC) United Nations Office on Drugs and Crime. (2012 December). Counter piracy programme: Support to the trial and related treatment of piracy suspects, issue ten. Retrieved from https://www.unodc.org/documents/easternafrica/piracy/CPP_brochure_December_2012.pdf

(UNODC) United Nations Office on Drugs and Crime. (2015). *Somali prison survey report: Piracy motivations and deterrents*. Retrieved from https://www.unodc.org/documents/Piracy/SomaliPrisonSurveyReport.pdf

van Wormer, K. S., & Bartollas, C. (2021). Gender-specific programming for female offenders. In K. van Wormer & C. Bartollas (Eds.), *Women and the criminal justice system* (pp. 68–95). Oxfordshire: Routledge.

Vargo, D. (2015). *Wild women of Boston: Mettle and moxie in the hub*. Mount Pleasant: Arcadia Publishing.

Widom, C. S., & Osborn, M. (2021). The cycle of violence: Abused and neglected girls to adult female offenders. *Feminist Criminology*, *16*(3), 266–285.

Young, A. J., & Valencia, M. J. (2003). Conflation of piracy and terrorism in Southeast Asia: Rectitude and utility. *Contemporary Southeast Asia: A Journal of International and Strategic Affairs*, *25*(2), 269–283.

Yuval-Davis, N. (1997). *Gender and nation*. London: Sage.

Conclusion

As I looked out the window of the small United Nations prop plane, the sandy Somali terrain seemed to rush up at me with a sense of urgency that conflicted with the slower pace of life in the Horn of Africa. Inside, lively chatter filled the air as everyone braced for the inevitably rough landing. Return flights from Somalia always seemed to be filled with a sense of excitement and lightness. Everyone seemed to be at ease with an unspoken sense of "we made it out safely". Most of us on the plane were foreigners, well aware that we were pushing our luck with each return trip to Somalia. At the time, kidnappings, assassinations, and bombings were a regular feature in a daily security briefing we received each morning at the UN compound before heading out to fulfill our mission tasks.

As the wheels touched down on the sun-scorched earth of the Baidoa airport, our bodies were violently jerked around in our seats for what felt like an unreasonably long time. You never get used to the rough landings. Each time, I recall my colleague preparing me for it on my first trip by describing it as feeling like you're going to crash and die but promising that I wouldn't. He was right. It was typical to have three to four stops on a flight from Somalia back to Nairobi. This was our third, and we had one more in Wajir just across the Kenyan border. As we deplaned, airport security awaited us to check our bags. The men made their way toward the two male security guards at the front of the plane, and the women headed toward the back of the plane to meet the two female security guards.

The presence of women security guards at Somalia airports was just one of the many changes I had noticed over time. When I first began working in Somalia, I was never checked at security, because it was *haram* for a male security guard to check a female. Now, the female guards patted me down so thoroughly it began to feel as if they were angry with me. All the men were cleared while three of us women continued to be searched. The Yemenese pilot came over and started harassing the Somalia women guards to hurry up. Like the rest of us, he was tired and ready to get back to the

DOI: 10.4324/9781003225201-6

respite of Nairobi. He eventually shoved past the women, grabbed our bags, and told us to get back on the plane.

The women security guards were part of a broader United Nations effort to bring women into the security and criminal justice sector throughout Somalia. During my tenure with the UNODC counter-piracy programming, we also helped place the first female prison guards in Hargeisa Prison. It was exciting to see these changes, but at the same time, I knew they could be pushed aside or delegitimized just as easily as the pilot had shooed off the women that day. Would the women be the first to be dismissed when the international donor money dries up? Or, would these women one day oversee entire wings full of female prisoners or patrol the streets of Somali towns under a mandate to root out female criminals?

Despite the important inroads made by feminist criminologists advocating to move women from the periphery to the center of crime and justice studies, those inroads have not come without their challenges. One of which, predicted in Adler's (1975) seminal text, was that second-wave feminism's achievement of social equality meant that women, ostensibly freed by second-wave feminism from the constraints of patriarchal norms, would break the law in unprecedented numbers. This sparked a real debate that continues today among feminists as to what "progress" in the context of women and the criminal justice system looks like. This raises several important questions that feminist scholars continue to engage with as they advance understandings of women and crime.

First, how should researchers reconcile the very real victimization so many women offenders suffer throughout their lives with the realities of the harms they cause others through their choices? The victim–offender overlap involving women has been well documented over a multitude of different crimes, including intimate partner violence (Tillyer & Wright, 2014), commercial sexual exploitation (Henderson & Rhodes, 2022), cyberbullying (Marcum et al., 2014), and gang membership (Pyrooz et al., 2014), to name a few. Is the appropriate response to implement victim services programming in conjunction with police enforcement (Bucerius et al., 2021) or to focus on early preventative nonviolent resocialization to minimize the risk of adult offending (Varlioglu & Hayes, 2022)? Further, emergent research suggests that finances, relationships, and addiction may be stronger determinants of female offending that past trauma and mental illness (Kruttschnitt et al., 2019). In turn, how should researchers reconcile this reality with the victim–offender overlap so common among men who break the law? Typical "gender-responsive" programs and approaches in jails and prisons only pertain to women, as if gender is a synonym for women. These programs target "high risk" young girls and women with women-specific needs assessments and programming (Walker et al., 2015, Duwe & Clark, 2015) while ignoring the gendered needs of men.

Second, does drawing attention to women's offending reinforce very old beliefs that position patriarchal control as in women's best interests due to our supposed irrationality, inability to control our emotions, and poor leadership skills (Lombroso & Ferrero, 1893/2004)? In other words, does studying women's offending, particularly among women offenders who appear to have other lawful alternatives to make a living, risk the message: "just look what happens when women think they can do whatever they want!". We see this unfolding in the contemporary movement to stop "modern day slavery" where prostitutes are blamed for creating the demand that fuels sex trafficking (Kempadoo, 2017). The solution is to conduct "raid and rescue" missions, many of which turn violent and arguably violate human rights, in the name of protecting potential trafficking victims (Jones & Edwards, 2018). Accordingly, women become the central focus of blame for the victimization of other women rather than holding the individuals who are actively doing the trafficking and exploitation accountable.

Third, does studying women offenders, particularly as leaders of organized criminal networks, contribute to the mass incarceration of women that began in the 1980s? As prosecutors in the United States expanded who could "count" as a drug offender by including longer prison sentences for a vast array of ancillary drug economy roles, prison populations grew including the number of women incarcerated. This trend is now being seen worldwide. Broadening *what* activities are criminalized also expands *who* can be considered a criminal, including women in roles that have traditionally been viewed as supportive and ancillary enough not to draw harsh punishment. What should feminist criminologist's role be in advocating for a balanced criminal justice approach to women involved in organized crime, and what should that balance entail? While ignoring women's participation in supportive roles and diminishing their agency and competency has been the historical norm, If we begin to focus on women's participation in crime and punish it accordingly, the result may be that more women are drawn into the criminal justice system. Perhaps, feminist criminologists should simultaneously lead an effort to retheorize crime and its role in relation to social and economic changes (van der Heijden & Schmidt, 2018; Carvalho et al., 2020).

Lastly, is the gendered socio-legal lens used to envision organized crime incorrectly focused on particular aspects of crime (i.e., hijacking ships vs. laundering money to finance a piracy operation) fundamentally wrong? Criminal justice systems worldwide tend to construct crimes involving the embodied potential of violence or aggression as a larger threat to individuals and societies than those that don't (Ilan & Sandberg, 2019). This tendency has elicited a movement within critical criminology to look "beyond criminology" to focus instead on social harms. The social harms approach allows for a broader investigation of harms that may occur across a lifespan rather than just with one criminal event (see also studies on slow violence: Nixon, 2011 and Ward, 2015) as well as consider states and organizations as perpetrators of violence (Hillyard et al., 2004). Scholars employing a social harm or zemiology approach may suggest that crimes perpetrated by women are often viewed as un-criminalized harms rather than crimes that warrant the full force of the criminal justice system, thus failing to consider multiple layers of victimization (Kotze & Boukli, 2016). Perhaps, then, focusing on both criminalized and un-criminalized harms can help move us past the gendered socio-legal construction of the criminal justice system and begin to deconstruct the victim/offender binary that often mirrors and replicates existing social inequalities (Kotze, 2018).

I have argued in this book for an alternative to the masculinized pirate mirage that recognizes and engages with piratical women as central, heterogeneous subjects of maritime piracy. Each chapter reflected on contradictions between dominant portrayals of piratical women and their lived experiences to challenge the male gaze of maritime fiction and reality and taught us something new about women's involvement in piracy. The popular culture figure of the woman pirates reveals beliefs regarding gender, sexuality, and race and the gendered glorification of crime in particular historical and cultural contexts. For example, in Victorian England only criminalized women could be considered sexy because society was conditioned to view sexuality as dangerous and scary. Women pirates were exotic and titillating and intentionally constructed in opposition to the domicile housewife. Centuries later, Disney dominates global film markets with racialized depictions of sexualized and dangerous piratical women of color. What does this say about contemporary American society? The label of piracy, and its pejorative undertones, have historically been applied to individuals believed to parasitically appropriate value they did not create through theft (Dawdy, 2011). Could these depictions be contemporary American society pushing back against the twenty-first-century socialist feminism that promotes intersectionality and inclusivity to protest gender oppression in the workplace, household, and community (Brenner, 2014)?

The gendered social organization of piracy reveals the nuances of broader social struggles within particular cultural and historical contexts. Historically, predominantly poor and working-class men who controlled piracy actively excluded women, replicating the class oppression that forced them into piracy, by promoting lies about women's supposedly inferior capabilities. Women were deemed unfit to survive the stressors and demands of maritime life and their mere presence onboard purportedly brought bad luck to any voyage. Accordingly, there are very few examples of women pirating at sea. Even today, most women's roles in piracy are relegated to shore-based activities that their male counterparts and broader society deem more (gender) appropriate. The diversity of these roles also reveals the complex means by which women exercise agency from a limited menu of life choices. Women such as Mary Read identified as a man to get work aboard a ship which led to a life of piracy. Several women held captive by Barbary pirates eventually married them to make a new life, albeit trading in one set of cultural restrictions for another. Even the fearless leaders of all-male crews like Grace O'Malley were navigating the boundaries of agency in a time before first-wave feminism when women, who law defined as property of their fathers and then their husbands, couldn't even buy property, get a divorce, choose who to marry, vote, or do anything, really.

Historical and practitioner accounts of piracy presented a limited understanding of the relationship between gender and crime in relation to piracy. Historians traditionally depict women pirates as "receiving" their skills from husbands, fathers, or male intimate partners as if those skills were genetically or sexually transmitted rather than cultivated by the women themselves. Similarly, practitioners' knowledge about women's roles in piracy is limited to the naval patrols, prison cells, and related data that is skewed by a socio-legal apparatus and military directives that prioritize the at sea and operational roles predominantly filled by men. In the case of Somalia, when women are found to be involved in piracy, regional governments step into redirect the focus on men to preserve the perceived virtue of the women. Perhaps, then it is more accurate to say that women may not necessarily be operating beyond the *gaze* of the law so much as they are operating beyond the *reach* of military and law enforcement. The gendered spaces of exception created by the resistance to policing women in Somalia presented opportunities for women to fill strategic shore-based piracy roles that intersected with well-established gendered expectations of social reproduction and carework. Whereas microsocial studies of Somali piracy reveal the intersections between gender norms and crime, the social construction of piracy and the legal and criminal justice responses continue to reflect macrosocial studies that depict women as victims who can't commit crimes, let alone be in charge of a criminal operation, perpetuating the academic knowledge vacuum surrounding women's role.

I concluded this manuscript while embarking on a new project designed to make my research more widely available to a general audience. After researching and publishing on the topic of Somali piracy for more than a decade, I still find it difficult to penetrate the popular piracy discourse grounded in quantitative datasets and military operations strategies. With every research trip back to East Africa I am reminded that there is so much more of the piracy story that needs to be told. As I comb through the piracy trial transcripts from Kenya and the Seychelles for my new project, a graphic novel on the Seychelles' first piracy trial, I can't help but reflect on some of the interesting older critiques of the International Criminal Tribunal for Rwanda and the International Criminal Court. Did the early twenty-first-century piracy trials of East Africa, facilitated by the UNODC and funded by Western donor countries, replicate an international criminal justice order that disproportionately targets black men as perpetrators? Over three hundred pirates were convicted—all were Somali men. The piracy labor of Somali women is invisible in the trials I am analyzing. This is particularly concerning, because the East African piracy trial transcripts, much like the historical trial transcripts from the Americas, England, and China, will serve as a main source of information in shaping how we come to *know* piracy, and pirates, of the early twenty-first century. We cannot allow the erasure of women in maritime crime to continue.

As feminist criminologists and political scientists continue to focus their energies on recentering women in studies of crime and justice, I encourage scholars to reflect on the complex realities of how this centering may impact legislative and policy approaches. What is the end goal? We need to continue to be mindful of what is productive, or counterproductive, in moving feminist goals of achieving rights and equality. I hope for a future in which crime events don't define individuals and one in which women all over the world have the freedom to choose their economic future in a way that doesn't make crime their most lucrative (or only) option.

References

Adler, F. (1975). *Sisters in crime: the rise of the new female criminal*. New York: McGraw-Hill.

Brenner, J. (2014). 21st century socialist-feminism. *Socialist Studies/Études Socialistes*, *10*(1), 31–49.

Bucerius, S. M., Jones, D. J., Kohl, A., & Haggerty, K. D. (2021). Addressing the victim–offender overlap: Advancing evidence-based research to better service criminally involved people with victimization histories. *Victims and Offenders*, *16*(1), 148–163.

Carvalho, H., Chamberlen, A., & Lewis, R. (2020). Punitiveness beyond criminal justice: Punishable and punitive subjects in an era of prevention, anti-migration and austerity. *British Journal of Criminology*, *60*(2), 265–284.

Dawdy, S. L. (2011). Why pirates are back. *Annual Review of Law and Social Science*, *7*, 361–385.

Duwe, G., & Clark, V. (2015). Importance of program integrity: Outcome evaluation of a gender-responsive, cognitive-behavioral program for female offenders. *Criminology and Public Policy*, *14*(2), 301–328.

Henderson, A. C., & Rhodes, S. M. (2022). "Got sold a dream and it turned into a nightmare": The victim-offender overlap in commercial sexual exploitation. *Journal of Human Trafficking*, *8*(1), 33–48.

Hillyard, P., Pantazis, C., Tombs, S., & Gordon, D. (2004). *Beyond criminology: Taking harm seriously*. Las Vegas: Pluto.

Ilan, J., & Sandberg, S. (2019). How 'gangsters' become jihadists: Bourdieu criminology and the crime-terror nexus. *European Journal of Criminology*, *16*(3), 278–294.

Jones, S., King, J., & Edwards, N. (2018). Human-trafficking prevention is not "sexy": Impact of the rescue industry on Thailand NGO programs and the need for a human rights approach. *Journal of Human Trafficking*, *4*(3), 231–255.

Kempadoo, K. (2017). Sex workers' rights organizations and anti-trafficking campaigns. In K. Kempadoo, J. Sanghera, & B. Pattanaik (Eds.), *Trafficking and prostitution reconsidered* (pp. 149–155). Oxfordshire: Routledge.

Kotzé, J. (2018). Criminology or zemiology? Yes, please! On the refusal of choice between false alternatives. In A. Boukli & J. Kotze (Eds.), *Zemiology* (pp. 85–106). Cham: Palgrave Macmillan.

Kotzé, J., & Boukli, A. (2016). Review of invisible crimes and social harms by Pamela Davies, Peter Francis and Tanya Wyatt. *British Journal of Criminology*, *56*, 818–820.

Kruttschnitt, C., Joosen, K., & Bijleveld, C. (2019). Research note: Re-examining the gender responsive approach to female offending and its basis in the pathways literature. *Journal of Offender Rehabilitation*, *58*(6), 485–499.

Lombroso, C., & Ferrero, G. (1893/2004). *Criminal woman, the prostitute, and the normal woman*. Duke University Press.

Marcum, C. D., Higgins, G. E., Freiburger, T. L., & Ricketts, M. L. (2014). Exploration of the cyberbullying victim/offender overlap by sex. *American Journal of Criminal Justice*, *39*(3), 538–548.

Nixon, R. (2011). *Slow violence and the environmentalism of the poor*. Cambridge, MA: Harvard University Press.

Pyrooz, D. C., Moule Jr., R. K., & Decker, S. H. (2014). The contribution of gang membership to the victim–offender overlap. *Journal of Research in Crime and Delinquency*, *51*(3), 315–348.

Tillyer, M. S., & Wright, E. M. (2014). Intimate partner violence and the victim-offender overlap. *Journal of Research in Crime and Delinquency*, *51*(1), 29–55.

van der Heijden, M., & Schmidt, A. (2018). Theorizing crime and gender in a long-term perspective. In E. Dermineur, A. Karlsson Sjogren, & V. Langum (Eds.), *Revisiting gender in European history, 1400–1800* (pp. 52–77). Oxfordshire: Routledge.

Varlioglu, R., & Hayes, B. E. (2022). Gender differences in the victim-offender overlap for dating violence: The role of early violent socialization. *Child Abuse and Neglect*, *123*, 105428.

Walker, S. C., Muno, A., & Sullivan-Colglazier, C. (2015). Principles in practice: A multistate study of gender-responsive reforms in the juvenile justice system. *Crime and Delinquency*, *61*(5), 742–766.

Ward, G. (2015). The slow violence of state organized race crime. *Theoretical Criminology*, *19*(3), 299–314.

Index

For Product Safety Concerns and Information please contact our EU
representative GPSR@taylorandfrancis.com
Taylor & Francis Verlag GmbH, Kaufingerstraße 24, 80331 München, Germany

www.ingramcontent.com/pod-product-compliance
Lightning Source LLC
Chambersburg PA
CBHW061335220326
41599CB00026B/5202
* 9 7 8 1 0 3 2 1 2 5 7 1 8 *